Natural sources of flavourings

Report No. 1

All rights reserved. No part of this publication may be reproduced, stored in a retrieval system, or transmitted, in any form or by any means, electronic, mechanical, photocopying, recording or otherwise, without the prior permission of Council of Europe Publishing.

Cover design: Graphic Design Workshop, Council of Europe
Layout: DTP Unit of the Council of Europe
Edited and published by: Council of Europe Publishing,
F-67075 Strasbourg Cedex

ISBN 92-871-4324-2
© Council of Europe, July 2000
Printed and bound in Germany by Koelblin-Fortuna-Druck

The Council of Europe

The Council of Europe was founded in 1949, in the wake of the second world war, in order to enhance European construction, based on a common set of values, contempt for which had brought about a time of strife and inhumanity.

The Organisation currently comprises 41 member states – in other words almost all the countries of the European continent, representing some 800 million Europeans. Since 1989, the Council has been one of the driving forces in bringing western, central and eastern Europe together. It works towards creating a common democratic and legal area structured around the European Convention on Human Rights. It develops many forms of co-operation on a wide range of issues to be addressed by our societies today, including education, social cohesion, protection of national minorities, the fight against all forms of intolerance, the prevention of crime and corruption, the consolidation of local democracy and the enhancement of Europe's cultural heritage.

Council of Europe Publishing is the official publisher of the Council of Europe, and reflects the many different aspects of the Council's work, addressing the challenges facing European society today. Our catalogue of over 1 200 titles in French and English includes topics ranging from international legal instruments and human rights to ethical and moral issues, society, environment, health, education and culture.

For more information please see our website: **http://book.coe.fr**

Note to the reader

For further information on *Natural sources of flavourings, Report No.1* or on the Council of Europe's work programme on flavouring substances, please contact:

Dr Peter Baum

Partial Agreement Department in the Social and Public Health field
Council of Europe
Avenue de l'Europe
F-67000 Strasbourg

Tel. +33 (0)3 88 41 21 76
E-mail: peter.baum@coe.int
Fax +33 (0)3 88 41 27 32

For a list of publications by Council of Europe Publishing on flavouring substances, please see at the back of the book.

For general information on the Council of Europe work programme on public health and consumer protection, please contact the Internet website: www.coe.fr/soc-sp

Table of contents

	Page
1. General introduction	7
2. Foreword	7
3. Definitions	8
4. Aims of the study	9
5. General principles	9
6. Classification system for natural flavouring sources and preparations	11
7. Active principles and other chemical components	13
8. List of natural sources of flavourings	13
Appendix 1: References	17
Appendix 2: List of "active principles" currently under evaluation by the committee	19
Appendix 3: List of "other chemical components" currently under evaluation by the committee of experts	21
Appendix 4: Definitions of types of natural flavouring preparations	23
Appendix 5: Preparation and format of datasheets for natural flavouring source materials	25
Appendix 6: Evaluated natural sources of flavourings	27
Alphabetical index of the natural sources of flavourings	267
List of national delegates of the Committee of Experts on Flavouring Substances	271
Council of Europe publications on flavouring substances	275

1. General introduction

Fields of activity of the Council of Europe

The competence of the Council of Europe is very wide and covers practically all aspects of European affairs, with the exception of defence matters. Where, however, a lesser number of states wish to engage in some action in which not all their European partners desire to join, they can conclude a "partial agreement" which is binding on themselves alone.

Partial Agreement in the Social and Public Health Field

It was on this basis that the Partial Agreement in the Social and Public Health Field was concluded in 1959 by the Council of Europe Committee of Ministers and revised in 1996 by the Committee of Ministers with effect from 1 January 1997. The following states are members of the Partial Agreement: Austria, Belgium, Cyprus, Denmark, Finland, France, Germany, Ireland, Italy, Luxembourg, the Netherlands, Norway, Portugal, Slovenia , Spain, Sweden, Switzerland and the United Kingdom of Great Britain and Northern Ireland.

The aim of the Partial Agreement public health activities is to protect the consumer from potential risks connected with the present-day way of life. The committees of experts provide the scientific base for national and international regulations concerning products which have a direct or indirect impact on the human food chain (control of foodstuffs, nutrition, food safety, consumer health, food contact materials, flavouring substances), pesticides, pharmaceuticals and cosmetics.

The steering committee responsible for the activities concerning health protection of the consumer is the Public Health Committee, composed of high-ranking Ministry of Health officials of the member states. It supervises the committees of experts' activities and specifies the general principles of the public health policy which these committees of experts have to follow in their work.

The Partial Agreement in the Public Health Field has always closely followed technical evolution and scientific progress and resolutions, guidelines and reports arising from its work are often of a pioneering nature.

In most fields, the approach is that of toxicological evaluation of chemical substance groups, the setting of technical and toxicological specifications, elaboration of maximum or guideline levels and preparation of inventory lists of chemical substances used by industry.

2. Foreword

Volume I of the 4th edition of the Council of Europe's "Blue Book" was published[1] in 1992 and dealt with chemically-defined flavouring substances. The Committee of Experts on Flavouring Substances, hereafter called committee of experts, is currently engaged in a major review of the safety-in-use of over 600 natural flavouring source materials, the conclusions of which will eventually be published in the second volume. This review builds on the Committee of experts' earlier evaluations of natural sources, the most recent of which were included in the 3rd edition of the "Blue Book" which was published[2] in 1981. In view of the time required to complete this large task, it has been agreed that interim reports should be published at regular intervals covering the source materials so far evaluated. This is the first of these reports and provides the Committee of experts' evaluations of 101 source materials.

The committee of experts is also reviewing the toxicity of certain potentially toxic chemically-defined substances known as "active principles" and "other chemical components" (see definitions in section 3) which occur in natural flavourings. The conclusions of this review will also form part of the second volume. A list of the "active principles" and "other chemical components" currently under evaluation by the committee of experts is given in Appendices 2 and 3.

The committee of experts has also considered the safety-in-use of smoke flavours, thermal process flavourings, flavourings produced by enzymatic and microbiological processes and flavourings produced using plant tissue culture and its conclusions in these areas have been published separately.[3-7]

3. Definitions

Natural sources of flavourings (or natural flavouring source materials) are materials of vegetable or animal origin, whether or not they are normally consumed as food, from which flavourings may be obtained. The Committee of experts has predominantly evaluated materials in the raw or dried state, with the exception of certain special products, such as vanilla, cocoa and black pepper, which are traditionally processed (e.g. fermented) before their use as source materials.

Flavouring preparations are products (other than chemically-defined flavouring substances), whether concentrated or not, with flavouring properties which are obtained from natural sources of flavourings by appropriate physical processes (including distillation and solvent extraction).[8] Definitions of the types of flavouring preparations included in this report, based on those specified by the International Organization for Standardization,[9] are given in Appendix 4.

Active principles are chemically-defined substances which occur in natural flavouring source materials and preparations and which, on the basis of existing toxicological data, should not be used as flavouring substances in their own right. They are therefore not included in the Council of Europe's list of chemically-defined flavourings.[1] Maximum limits in foodstuffs as consumed should be set on the basis of existing toxicological data or with reference to a Tolerable Daily Intake (TDI).[a]

Other chemical components are chemically-defined substances which occur at high levels in certain natural flavouring source materials and preparations. The Committee of experts has evaluated these substances and considered them to be of less toxicological concern and hence not to be covered by the definition of "active principle". These substances may be used as chemically-defined flavouring substances in their own right, but maximum limits in foodstuffs as consumed may be set on the basis of existing toxicological data or with reference to a TDI. Substances which have been evaluated by JECFA,[b] and for which it has set an ADI, are not included in the Committee of experts' list of "other chemical components".

---

a. Acceptable Daily Intake (ADI) was originally defined by the World Health Organization as "an estimate by JECFA[b] of the amount of a food additive, expressed on a body weight basis, that can be ingested daily over a lifetime without appreciable health risk". The definition of Tolerable Daily Intake (TDI) is the same as that for ADI but is generally used only for substances classed as contaminants. The Committee of experts has extended the definition of TDI to cover "active principles" and "other chemical components".
b. JECFA: Joint FAO/WHO Expert Committee on Food Additives.

4. Aims of the study

The Committee of experts set itself the following aims in its study of natural sources of flavourings and preparations thereof:

– to draw up a list of natural sources of flavourings and preparations, indicating their acceptability for use in food;

– to draw attention to certain natural sources of flavourings and preparations which present a hazard to public health;

– to draw up a list of "active principles" and set maximum limits for such substances in foodstuffs;

– to draw up a list of "other chemical components" and, where appropriate, to set maximum limits for such substances in foodstuffs;

– to supplement and revise the lists referred to above; and

– to recommend to manufacturers of flavouring source materials and flavouring preparations the information required by the Committee of experts to complete its evaluations of their safety-in-use.

5. General principles

5.1 Considerations

The committee of experts' review only covers source materials for which interested manufacturers have requested a safety evaluation; it is therefore not exhaustive. The committee of experts' recommendations on the safety-in-use of natural source materials and preparations as flavourings should not be interpreted as approval for their use in any other application.

The committee of experts is conscious that the level of use of natural flavouring source materials and preparations is generally low and their use is often limited by their organoleptic properties. In general, they should be used at the lowest level which is technologically effective. However, these factors are not in themselves sufficient to guarantee that there are no public health risks associated with their use.

The committee of experts recognises that up-to-date information on the current usage of natural source materials and preparations is an important part of the evaluation process. As part of this process, the committee of experts receives information from the International Organization of the Flavor Industry (IOFI) on the current levels of use of source materials and preparations in foods and beverages. Where necessary, the committee of experts may also ask IOFI to provide information on the chemical composition of source materials and preparations and on any relevant toxicological studies. Since 1995, IOFI has been invited to attend part of each committee of experts' meeting, making the exchange of information faster and more efficient.

The evaluation of natural flavouring source materials and preparations is a large and complex task. Several parts (e.g. fruit, leaves, roots) of the same source species can be used in natural flavourings and, depending on the species, these can be added as such, or with very limited processing, such as crushing, grinding or drying (e.g. herbs and spices), and/or they can be used to derive natural flavouring preparations. It is also possible for several different types of flavouring preparations (e.g. essential oils, extracts, etc.) to be prepared from the same part. In addition, there can be significant qualitative and quantitative differences in chemical composition between different parts and between different preparations. For example, alkaloids are found in the bark but not in the fruit of pomegranate, while anthraquinones are present in the leaves of rhubarb but are found at only trace levels in the stalks.

In the 3rd edition of the "Blue Book",[2] one overall classification was given for each natural flavouring source. This approach has been refined in the current review. Ideally, all the different parts of the plant, and any preparations derived from them which are used in flavourings should be classified separately. However, although they are likely to have closely related chemical compositions, full details are not always available for each one. In addition, a very long list of classifications would be needed. As a compromise, the committee of experts has, in most cases, given a separate classification for each part of the source species used directly as such and a group classification for all the preparations derived from each part.

Natural flavouring source materials and preparations may contain substances defined by the committee of experts as "active principles" or "other chemical components" (see definitions in Section 3). The addition of these natural flavourings to foods and beverages will obviously result in the presence of these substances in the products concerned. In setting limits for these substances in foods and beverages, the committee of experts considers that the general limits should be set as low as technologically feasible to restrict consumer exposure. Where the "active principle" concerned is a potential genotoxic carcinogen, the general limit is set at the limit of detection. The committee of experts recognises, however, that there may be a small number of traditional foods and beverages in which the levels of these substances could be higher than the general limits. The committee of experts therefore sets higher limits for these few tightly-defined food categories.

Natural flavourings are prepared from vegetable or animal sources which may themselves be consumed as part of the human diet. The committee of experts evaluates the safety-in-use of the flavourings derived from these foods and not the foodstuffs themselves. Only foods used by the flavouring industry to derive natural flavourings are listed. Similarly, if only one part (e.g. leaves) of a source is used in flavourings, then other parts of the source are not listed.

The consumption of natural flavouring source materials and preparations listed in this report, as well as many other food ingredients, may sometimes sensitise or produce hypersensitivity reactions in some people. At present, such reactions cannot reliably be predicted from tests on laboratory animals or from *in vitro* studies. Therefore, in common with other international organisations, the committee of experts has not been able to consider these issues in arriving at its decisions.

When evaluating natural flavouring source materials and preparations, the committee of experts has often had to take decisions based on very limited toxicological data available from published literature or other sources. Hence, it has had to adopt less rigorous criteria than those generally applied in the safety evaluation of food additives. The value of the available data has been judged on a case-by-case basis rather than by adhering to rigid requirements. A general description of the criteria for evaluation is given below.

5.2 Toxicological criteria for classification

In general, 28-day studies in rats and in vitro mutagenicity studies have been regarded as the minimal requirement for inclusion of a part of a source species, or a flavouring preparation, in a fully acceptable category. However, this requirement has frequently been reduced, particularly in the following situations:

> The part of the source species used in natural flavourings is commonly consumed as a food in Europe. The division of sources into food and non-food has been made on the basis of the committee of experts' experience of the European food supply. The committee of experts is aware of a number of plant foods which are safely consumed in other continents but this may only be because their populations have developed ways of removing or inactivating the toxins they contain. Without knowing exactly how each food consumed outside Europe is normally handled, the amount which might normally be consumed and the likelihood that any adverse effects would have been noted, it is not possible to take this history of use into account in the evaluation of such source materials.
>
> The chemical composition of the flavouring source material or preparation is closely related to the chemical composition of another flavouring source or preparation which has been more thoroughly studied.
>
> The major constituents of the source materials and/or preparations are thought unlikely to cause adverse effects at the current level of use, for example because they have previously been evaluated by the committee of experts or by the EU's Scientific Committee on Food or another recognised international expert group dealing with flavourings and found to be fully acceptable as flavouring substances.
>
> Conversely, the requirement for toxicological data has been increased in certain situations such as where the data available on source materials and preparations, or their constituents, have given cause for concern and further studies are needed for reassurance.

The classification of parts and/or preparations also takes into account the presence of "active principles" or "other chemical components" in the source species. This is discussed in more detail in Sections 6 and 7.

6. Classification system for natural flavouring sources and preparations

In the 3rd edition of the "Blue Book",[2] natural flavouring sources were listed and classified into four categories (N1-N4). This approach has been refined by considering individual parts and preparations and allocating them a single classification or, in the case of preparations derived from the same part, a group classification. Flavouring source materials and preparations have been classified into six numbered categories as follows:

Category 1

Plants, animals and other organisms, and parts of these or products thereof, normally consumed as food items, herbs or spices in Europe for which it is considered that there should be no restrictions on use.

Flavouring preparations, which are not themselves consumed as food but which are derived from plants, animals and other organisms and parts of these or products thereof, normally consumed as food items, herbs or spices in Europe. These preparations, on the basis of the information available, are not considered a risk to health in the quantities used.

Category 2

Plants, animals and other organisms, and parts of these or products thereof, and preparations derived therefrom, not normally consumed as food items, herbs or spices in Europe.

These source materials and preparations, on the basis of the information available, are not considered to constitute a risk to health in the quantities used.

Category 3

Plants, animals and other organisms, and parts of these or products thereof, normally consumed as food items, herbs or spices in Europe which contain defined "active principles" or "other chemical components" requiring limits on use levels.

Flavouring preparations, which are not themselves consumed as food but which are derived from plants, animals and other organisms and parts of these or products thereof, normally consumed as food items, herbs or spices in Europe which contain defined "active principles" or "other chemical components" requiring limits on use levels.

These source materials and preparations are not considered to constitute a risk to health in the quantities used provided that the limits set for the "active principles" or the "other chemical components" are not exceeded.

Category 4

Plants, animals and other organisms, and parts of these or products thereof, and preparations derived therefrom, not normally consumed as food items, herbs or spices in Europe, which contain defined "active principles" or "other chemical components" requiring limits on use levels.

These source materials and preparations are not considered to constitute a risk to health in the quantities used provided that the limits set for the "active principles" or the "other chemical components" are not exceeded.

Category 5

Plants, animals and other organisms, and parts of these or products thereof, and preparations derived therefrom, for which additional toxicological and/or chemical information is required.

These could temporarily be acceptable provided that any limits set for the "active principles" or the "other chemical components" are not exceeded.

Category 6

Plants, animals and other organisms, and parts of these or products thereof, and preparations derived therefrom, which are considered to be unfit for human consumption in any amount.

As described in Section 5.1 iv., all preparations derived from each part of a source species have generally been given a group classification, but it must be stressed that this classification applies only to the preparations listed on each datasheet. Other preparations may have different compositions and these would have to be evaluated separately before being included in the existing classification.

Source materials and preparations classified in categories 1 and 2, from food and non-food sources respectively, are considered to be fully acceptable, and no limits are required on their use.

Those in categories 3 and 4 are also fully acceptable, provided that their use does not result in unacceptably high intakes of "active principles" or "other chemical components".

Where flavouring source materials and preparations in categories 3 and 4 contain an "active principle", a maximum limit has been recommended for the level of the "active principle" in foodstuffs in which the flavourings are used. The setting of such limits does not necessarily imply that the use of a particular flavouring source material or preparation must be restricted, since the amounts added to food may not be high enough to exceed the limit. Other source materials and preparations in categories 3 and 4 may contain "other chemical components" for which maximum limits in foodstuffs as consumed have been set on the basis of existing toxicological data or with reference to a TDI.

Category 5 contains source materials which have not been fully evaluated due to a lack of data, but which have been categorised as temporarily acceptable on the basis of the available information. In some cases there is insufficient information on the composition of the source material or its flavouring preparations, while in others it has not been possible to assess the significance of particular components which are known to occur.

Source materials placed in category 6 are considered to be unsuitable for use as sources of food flavourings due to the presence of particularly toxic substances.

Natural flavouring source materials whose evaluation has been postponed since there is no evidence that they are currently in use in Europe have not been categorised.

7. "Active principles" and "other chemical components"

As mentioned in Section 5.2 ii., "active principles" are being reviewed separately to facilitate the evaluation of the flavouring source materials and preparations in which they occur. Previous editions of the "Blue Book" contained only a summary list of the limits recommended for each "active principle" in foods and beverages. In the 4th edition, datasheets will be included for all "active principles". These will summarise the toxicological evidence for the classification of the substance as an "active principle", explain the derivation of the TDI (if one has been set) and outline any further data which may be required. They will also list the limits set for each of the "active principles" in foods and beverages. A list of the substances currently classified as "active principles", together with a list of those substances which were earlier considered to fall in this category, are given in Appendix 2 to this report. A number of "other chemical components" are also being evaluated separately and are listed in Appendix 3. Finalised datasheets for both "active principles" and "other chemical components" will be published in due course.

8. List of natural sources of flavourings

In this and future interim reports, information on those source materials which have been evaluated is provided in three sections:

– materials which have been evaluated and found acceptable, or temporarily acceptable, for use as sources of food flavourings;

– materials which are considered to present a hazard to human health and which should not be used as source materials; and

- materials whose evaluation has been postponed as there is no evidence that they are currently in use as source materials.

In common with the lists published in previous editions, the natural flavouring sources in the first and second categories are listed by Council of Europe (CE) number (provisional) and preferred systematic (scientific Latin) species name, as well as by order and family. However, unlike the earlier editions, the sources are ordered alphabetically by systematic name rather than by CE number.

Each source species is identified primarily by its systematic name based on the botanical nomenclature of Zander[10] or, where the source species is not listed in that reference book, using that of Mansfeld.[11] Source species not listed in either book are treated on a case-by-case basis, e.g. based on Hager,[12] Rehm[13] or Hoppe.[14] Since there have been many instances where species have been reclassified and renamed and where the same species is known by different scientific names, there is an entry for synonyms given or previous names by which the organism may be known. Common names (in English, French, German, Italian and Spanish) are also given, but it should be noted that a name used in one region to refer to a particular plant may be used elsewhere to refer to another quite unrelated species. Hence these names may not be as reliable as the systematic names.

Many of the flavouring sources which are widely used have been evaluated or listed previously, as indicated by a Steinmetz number and/or a Flavor and Extract Manufacturers' Association (FEMA) number (discussed further below). It should be noted that CE (and Steinmetz) numbers refer to source species, while FEMA numbers refer to particular preparations derived from specified parts of a source species. Hence, a single CE number may cover several FEMA numbers.

In previous editions, the information on natural flavouring sources considered acceptable, or temporarily acceptable, for use in flavourings was limited to the name of each source species, along with a list of the parts used as flavouring sources and their classifications. This has been expanded in the current edition to provide an overview of all the relevant information available on each source in the form of a standard datasheet.

Each datasheet provides a list of the "important constituents", including the known "active principles" and "other chemical components", found in each part of the plant and/or preparation used in flavourings. Where possible, the levels at which the various constituents are present in the parts/preparations are given. However, for some source materials, only qualitative information is currently available. Differences also exist in the amount of detail provided, with some datasheets containing information on a large number of substances and others listing only a few key constituents. These differences reflect not only the varying amount of published information currently available on individual source materials, but also the alternative approaches taken by delegations to preparing this section of the datasheet. For example, some delegations have described all constituents found at concentrations below a certain "threshold value" as "trace" or have omitted them altogether, while others have included all "important constituents" whatever their concentration.

The committee of experts is currently discussing the question of thresholds in relation to "important constituents" and has set up an ad hoc group to consider the issue. The group will also consider whether a threshold should be used in the classification of source materials and preparations which contain "active principles" and "other chemical components". In the meantime, the committee of experts has agreed that in this report, all source materials and preparations should be classified as category 3 or 4 whatever the concentrations of "active principles" and "other chemical components" present.

Several key reference books have been consulted in preparing the datasheets, including publications by Arctander,[15] Duke,[16] Fenaroli,[17] Leung,[18] the UK Ministry of Agriculture, Fisheries and Food (MAFF),[19] Opdyke,[20] Tanaka[21] and Usher.[22] In order to save space in the datasheets, these references are simply listed by author name.

As mentioned above, many of the flavouring sources covered by this report have been the subject of national and international evaluations, and these are noted in the datasheets. For example, the UK Food Additives and Contaminants Committee (FACC) carried out an extensive review of over 500 flavouring source materials in the 1970s, the results of which were published in 1976.[23] In the US, since 1958, the Food and Drug Administration (FDA), with the assistance of FEMA, has been compiling lists of flavouring materials considered to be GRAS (Generally Recognised As Safe) and these are included in Chapter 21 of the Code of Federal Regulations (CFR). In addition, the FEMA expert panel is continuing to publish separate lists of materials which it considers to be GRAS.

Further information on the preparation and format of natural flavouring source material datasheets is given in Appendix 5. A list of materials whose evaluation has been postponed, as there is no evidence that they are currently in use as flavouring source materials, is given in Appendix 6.

Appendix 1

References

1. Council of Europe (1992). *Flavouring Substances and Natural Sources of Flavourings. 4th Edition. Volume I. Chemically-defined Flavouring Substances.* ISBN 2-7160-0147-2.

2. Council of Europe (1981) *Flavouring Substances and Natural Sources of Flavourings. 3rd Edition.* ISBN 2-7160-0081-6. [out of stock]

3. Council of Europe (1992). *Council of Europe Guidelines concerning the transmission of flavour of smoke to food,* Health Protection of the Consumer Series. Council of Europe Press, Strasbourg. ISBN 92-871-2191-5.

4. Council of Europe (1992). *Health aspects of using smoke flavours as food ingredients,* Health Protection of the Consumer Series. Council of Europe Press, Strasbourg. ISBN 92-871-2189-3.

5. Council of Europe (1994). *Council of Europe Guidelines for flavouring preparations produced by enzymatic or microbiological processes,* Health Protection of the Consumer Series. Council of Europe Press, Strasbourg. ISBN 92-871-2586-4.

6. Council of Europe (1995). *Guidelines for safety evaluation of thermal process flavourings,* Health Protection of the Consumer Series. Council of Europe Publishing, Strasbourg. ISBN 92-871-2811-1.

7. Council of Europe (1998). *Council of Europe Guidelines for flavouring preparations produced by plant tissue culture,* Health Protection of the Consumer Series. Council of Europe Publishing, Strasbourg. ISBN 92-871-3738-2.

8. European Community (1988). Council Directive of 22 June 1988 on the approximation of the laws of Member States relating to flavourings for use in foodstuffs and to source materials for their production (88/388/EEC). *Official Journal of the European Communities* No.. L 184/61-67.

9. International Organization for Standardization (1997). *Aromatic natural raw materials – Vocabulary.* ISO 9235: 1997. International Organization for Standardization, Geneva.

10. Zander, R. (1984). *Handwörterbuch der Pflanzennamen,* 13th Edition. F. Encke, G. Buchheim, S. Seybold (Eds.). Verlag Eugen Ulmer, Stuttgart.

11. Mansfeld, R. (1986). *Verzeichnis landwirtschaftlicher und gärtnerischer Kulturpflanzen (ohne Zierpflanzen),* Volumes 1-4, 2nd Edition. J. Herausgegeben von Schulte-Motel (Ed.). Springer-Verlag, Berlin, Heidelberg & New York.

12. List, P.H. and Hörhammer, L. (1967-80). *Hager's Handbuch der Pharmazeutischen Praxis,* Volumes 1-8, 4th Edition. Springer-Verlag, Berlin, Heidelberg & New York.

13. Rehm, S. (Ed.) (1994). *Multilingual Dictionary of Agronomic Plants.* Kluwer Academic Publishers, Dordecht.

14. Hoppe, H.A. (1977). Drogenkunde, Vol. 1 *Angiospermen,* Vol. 2: *Gymnospermen, Kryptogamen, Tierische Drogen.* Walter de Gruyter, Berlin & New York.

15. Arctander, S. (1960). *Perfume and Flavor Materials of Natural Origin.* Arctander, Elizabeth, New Jersey, USA.

16. Duke, J.A. (1985). *Handbook of Medicinal Herbs.* CRC Press, Boca Raton, Ann Arbor, London & Tokyo.

17. Burdock, G.A. (Ed.) (1995). *Fenaroli's Handbook of Flavor Ingredients.* Volume I and II. 3rd Edition. CRC Press, Boca Raton, Ann Arbor, London & Tokyo.

18. Leung, A.Y. and Foster, S. (1996). *Encyclopedia of Common Natural Ingredients used in Food, Drugs and Cosmetics.* 2nd Edition. John Wiley & Sons , Inc., New York, Chichester, Brisbane, Toronto & Singapore.

19. UK Ministry of Agriculture, Fisheries and Food. (1995). Flavourings in Food. *Food Surveillance Report* No. 48. HMSO, London.

20. Opdyke, D.L.J. (1979). Monographs on Fragrance Raw Materials. *Food and Cosmetics Toxicology,* 17, Special Issue V. Pergamon Press, Oxford, London, New York, Paris.

21. Tanaka, T. (1976). *Tanaka's Cyclopedia of Edible Plants of the World.* S. Nakao (Ed.). Keigaku Publishing Co., Tokyo.

22. Usher, D. (1974). *Dictionary of Plants Used by Man.* Constable, London.

23. UK Food Additives and Contaminants Committee. (1976). *Report on the Review of Flavourings in Food.* FAC/REP/22. HMSO, London.

Appendix 2

List of "active principles" currently under evaluation by the committee

"Active principle"	Status of evaluation	TDI	Limits
Agaric acid	Not yet started	-	-
Aloin	Under evaluation	-	To be agreed
b-Asarone	Complete	NA[1]	Food: 0.05 mg/kg[2] Beverages: 0.05 mg/kg[2] Exceptions: Alcoholic beverages (traditionally flavoured with calamus): 0.5
Berberine	Not yet started	-	-
Capsaicin	Complete	0.2 mg/kg bw/day	Food: 5 mg/kg Beverages: 5 mg/kg Hot food and beverages: 10 mg/kg Exceptions: Hot ketchup: 20 mg/kg Tabasco, harissa, hot pimento oils and similar preparations: 50
Coumarin	Under evaluation	NA[1]	To be agreed
Dill apiol	Complete	NA[1]	None set[3]
Elemicin	Complete	NA[1]	None set[3]
Estragole	Under evaluation	NA[1]	To be agreed
Furocoumarins	Under evaluation	-	-
Geniposide	Not yet started	-	-
Glycoalkaloids	Under evaluation	-	-
Glycyrrhizic acid	Not yet started	-	-
Hydrocyanic acid	Complete	0,02 mg CN-/kg bw/day	Food: 0.5 mg/kg Beverages: 0.05 mg/kg Exceptions: Stone fruit juices: 0.5 mg/kg Canned stone fruit and stone fruit preserves and purees: 2 mg/kg Marzipan, nougat and other similar products: 50 mg/kg Almond- and/or marzipan (or other similar product)- containing confectionery and baked goods: 10 mg/kg "Special" almond and/or marzipan (or other similar products)-containing confectionery and baked goods e.g. "amaretti", "Dresdner Christstollen", "schwarzbrötchen",

"Active principle"	Status of evaluation	TDI	Limits
			chocolate enrobed marzipan, marzipan novelties: 40 mg/kg Alcoholic beverages, for every 1% alcohol by volume: 0.5 mg/kg
Hypericin	Complete	0.0003 mg/kg bw/day	Food: None Beverages: 0.4 mg/kg
Menthofuran	Complete	0.1 mg/kg bw/day (joint TDI with pulegone)	Food: 20 mg/kg Beverages: 20 mg/kg Mint/peppermint flavoured alcoholic beverages: 100 mg/kg Mint/peppermint flavoured confectionery: 200 mg/kg Mint/peppermint flavoured chewing gum: 1000 mg/kg
Methyleugenol	Under evaluation	NA[1]	To be agreed
Myristicin	Complete	NA[1]	None set[3]
Parsley apiol	Complete	NA[1]	None set[3]
Polyacetylene compounds	Not yet started	-	-
Pulegone	Complete	0.1 mg/kg bw/day (joint TDI with menthofuran)	Food: 20 mg/kg Beverages: 20 mg/kg Mint/peppermint flavoured alcoholic beverages: 100 mg/kg Mint/peppermint flavoured confectionery: 100 mg/kg Intensely strong mint/peppermint flavoured confectionery: 200 mg/kg Mint/peppermint flavoured chewing gum: 350 mg/kg
Pyrrolizidine alkaloids	Not yet started	-	-
Safrole	Under evaluation	NA[1]	To be agreed
Thujone	Under evaluation	0,01 mg/kg bw/day	To be agreed
Xanthones	Under evaluation	-	-

Notes:
1 NA= Not Attributed. Considered as a potential genotoxic carcinogen, therefore no TDI set.
2 The limit has been set on the basis of the most sensitive method at this time. A lower limit of determination has been established in some laboratories and it is hoped that such methods will be further developed and be more widely available. If this is the case, these limits will be revised.
3 The Committee considers that limits are not required for myristicin and elemicin as their intakes (mainly from nutmeg) will be restricted via a limit for safrole in nutmeg-containing foods. Intakes of dill and parsley apiol are already very low.
The following substances have been deleted from the Committee's list of "active principles": allyl isothiocyanate and quinine (see list of "other chemical components"); trans-anethole (JECFA ADI); cocaine; quassine; and santonin.

Appendix 3

List of "other chemical components" currently under evaluation by the committee of experts

"Other chemical component"	Status of evaluation	TDI	Limits
Allyl isothiocyanate	Complete	0.06 mg/kg bw/day	None set
Caffeine	Under evaluation	-	To be agreed
Camphor	Under evaluation	To be agreed	To be agreed
Carvacrol	Complete	2.5 mg/kg bw/day	Food: 5 Beverages: 2 Exceptions: Confectionery: 25
Eucalyptol	Complete	0,2 mg/kg bw/day (provisional)	To be agreed
Parasorbic acid	Under evaluation	To be agreed	To be agreed
Quinine	Complete	Not established	Food: None Beverages (soft drinks): 100 mg/l

Notes: The Committee has agreed that substances which have been evaluated by JECFA, and for which it has set an ADI, should not be included in the Committee's list of "other chemical components". The following substances have therefore been deleted from this list: citral, eugenol, limonone, menthol, and methyl salicylate.

Appendix 4

Definitions of types of natural flavouring preparations

Definitions of the types of natural flavouring preparations included in this report, based on those specified by the International Organization for Standardization,7 are given below.

1. *Raw materials*

   Exudate: natural raw material excreted by plants either spontaneously or after wounding.

   Natural oleoresin: exudate consisting mainly of volatile and resinous constituents (e.g. pine oleoresin, gurjum).

   Balsam: natural oleoresin characterised by the presence of benzoic and/or cinnamic derivatives (e.g. Peru balsam, Tolu balsam, benzoin, styrax).

   Gum: exudate consisting mainly of polysaccharides (e.g. gum arabic, tragacanth gum).

   Gum resin: exudate consisting mainly of resinous constituents and gums (e.g. shellac gum).

   Gum oleoresin: exudate consisting mainly of resinous constituents, gums and certain amounts of volatile constituents (e.g. myrrh, olibanum, opoponax, galbanum).

2. *Derived products: resinous materials*

   Resin: product obtained from natural oleoresins by removing, as far as possible, the volatile constituents (e.g. rosin).

3. *Derived products: volatile products*

   Essential oil: product obtained from vegetable raw material:
   - by distillation with water or steam;
   - from the epicarp of Citrus fruits by a mechanical process; or
   - by dry distillation.

   (NB: The essential oil is subsequently separated from the aqueous phase by physical means.)

   Essential oil obtained by steam distillation: essential oil obtained by distillation with or without water in a still (e.g. pepper oil (with water), lavender oil (without water)).

   Cold-pressed essential oil: essential oil obtained from the epicarp of *Citrus* fruits by mechanical means at room temperature.

   Essence oil: essential oil obtained from fruit juices during concentration or UHT (flash pasteurisation) treatment

   Rectified essential oil: essential oil which has been subjected to fractional distillation in order to modify the content of certain constituents (e.g. mint essential oils).

   "Terpene-less" essential oil: essential oil from which the monoterpenic hydrocarbons have been mainly removed.

   "Terpene- and sesquiterpene-less" essential oil: essential oil from which the mono-and sesquiterpenic hydrocarbons have been mainly removed.

"X-less" essential oil: essential oil from which the component "x" has been partly or completely removed (e.g. essential oil of bergamot with partially reduced bergapten content; essential oil of Mentha arvensis with partially reduced menthol content).

Folded oil/concentrated oil: essential oil which has been processed to concentrate the components of interest by physical means.

Dry-distilled oil: essential oil obtained by dry distillation of woods, barks or roots without added water or steam (e.g. essential oil of cade (Juniperus oxycedrus), essential oil of the bark of the birch tree).

Volatile concentrate: concentrated water-soluble volatile substance recovered from the evaporated water of fruit or vegetable juices.

Distillate: product of condensation obtained after distillation of a natural raw material.

Alcoholate: distillate which results from the distillation of a natural raw material in the presence of ethanol at variable concentrations

Aromatic water: aqueous distillate remaining after steam distillation when the essential oil has been separated.

Terpenes: products mainly consisting of hydrocarbons obtained as by-products from an essential oil by concentration or distillation, or other isolation techniques.

4. *Derived products: extraction products*

Tincture: solution obtained by maceration of a natural raw material in the presence of ethanol at variable concentrations (e.g. tincture of benzoin, tincture of ambergris).

Extract: product obtained by treating a natural raw material with a solvent then, after filtration, removal of the solvent by distillation, except in the case of use of a non-volatile solvent.

Concrete: extract with a characteristic odour, obtained from a fresh vegetable raw material by extraction with a non-aqueous solvent.

Resinoid: extract with a characteristic odour, obtained from a dried vegetable raw material by extraction with a non-aqueous solvent.

Absolute: product with odour, obtained from a concrete or a resinoid by extraction with ethanol at room temperature.
(NB: The ethanolic solution is generally cooled and filtered in order to remove the "waxes"; the ethanol is then removed by distillation.)

Oleoresin: extract of spices or aromatic herbs with a characteristics odour and/or flavour (e.g. pepper oleoresin, ginger oleoresin).

Non-concentrated extract/single-fold extract: product obtained by treating a natural raw material with a non-removed solvent (e.g. cocoa nibs in propylene glycol, asafoetida in peanut oil, vanilla in ethanol).

Appendix 5

Preparation and format of datasheets for natural flavouring source materials

Datasheets are prepared for all the natural flavouring source materials which are being evaluated by the committee of experts, with responsibility for their preparation being divided between delegations. To ensure consistency in the presentation of the data, a standard format has been agreed as follows:

SYS name	Appears as the heading of each data sheet
CE No.	
Steinmetz No.	Reference number in Steinmetz Codex Vegetabilis (1957 edition)
FEMA No.	*"preparation or part number"* e.g. Valerian root extract:3009; Valerian root oil:3100
Order	
Family	
Name	E F D I SP
Synonyms	i.e. Latin names. Synonyms in English, French, German, Italian of Spanish are noted under each respective language
Parts used	Parts of plant normally used in flavourings
Important constituents	Components should be classified according to their chemical structure. Levels at which the constituents are present in each part of the plant or preparation should be given where available
Active principles	If there is no knowledge of any active principles present, the following phrase is used: "Not known"
Other chemical components	If none present, use: "Not known"
Products in which used	The main food categories in which parts/preparations are used should be listed. Pharmacological and cosmetic preparations and uses should be excluded
Level of use	An indication of the amount of preparation, from each part, which is used in different types of food. If no information, use: "No information available"
Preparation	The types of preparation made from each part listed in section 2 should be stated, e.g. oleoresin, extract, etc. If no preparations, use: "None"
Main toxicological data	Data on the following should be included if available: metabolism; sub-acute and sub-chronic toxicity; chronic toxicity; carcinogenicity;

	reproductive and teratogenicity studies; mutagenicity; other relevant studies, e.g. photosensitivity, beneficial effects. The part of the plant or the type of preparation which has been tested should be indicated in each case. Toxicity data on main constituents may also be included. Mention should be made if any of the major constituents are nature-identical flavouring substances evaluated in volume 1 of the 4th edition of the Blue Book, e.g. vanillin. If no data, use: "No data available". LD_{50} values should not be used
Data needed	If no data needed, use: "No data required". Standard phrase for data needed, use (relevant parts of): "Chemical composition and, if necessary, 28-day oral study and mutagenicity studies on (*relevant part/s and/or preparation/s*)"
Specific observations	Explanatory remarks to clarify e.g. classification, if relevant
Classification and limits	Only parts/preparations used should be classified. If an active principle or other chemical component is present: "With limits on *name of active principle/other chemical component*"
National/int. evaluation	Any evaluations carried out, e.g. by UK FACC (1976), FDA and FEMA.
Main references	References should be listed as follows;
	Journals: International abbreviation of scientific publication, No of journal:pages (Year of publication)
	Books: Author and/or editor, Title, Publisher, Pages if relevant (Year of publication)
Data bases used	List as follows:
	Database (from year-until year)
	"*Keywords:* ..."

Appendix 6

Evaluated natural sources of flavourings

Revised datasheets

CoE No.	Latin name	Active principles (AP)/ Other chemical components (CC)
1	Abelmoschus moschatus Moench.	Not known
11	Acer saccharum Marsh.	Not known
12	Achillea millefolium L.	AP Thujone CC Camphor, eucalyptol
15	Aframomun melegueta K. Schum.	Not known
19	Agropyron repens (L.) Beauv.	CC Carvacrol
264	Aloysia triphylla (L'Herit.) Britt.	Not known
33	Amyris balsamifera L.	Not known
56	Angelica archangelica L.	AP Coumarin, furocoumarins
2008	Annona cherimola Mill.	AP Hydrocyanic acid CC Camphor, eucalyptol,
46	Annona squamosa L.	AP Hydrocyanic acid CC Eucalyptol
49	Anthoxanthum odoratum L.	AP Coumarin
50	Anthriscus cerefolium (L.) Hoffm.	AP Estragole
60	Artemisia abrotanum L.	AP Methyleugenol, thujone CC Eucalyptol
61	Artemisia absinthium L.	AP Thujone CC Camphor
64	Artemisia dracunculus L.	AP Elimicin, estragole, methyleugenol, polycetylene compounds, thujone
66	Artemisia spicata Wulf.	AP Thujone CC Eucalyptol
68	Artemisia umbelliformis Lam.	AP Thujone
69	Artemisia pallens	Not known
70	Artemisia pontica L.	AP Thujone CC Eucalyptol
71	Artemisia vallesiaca Lam.	AP Estragole CC Camphor, eucalyptol
72	Artemisia vulgaris L.	AP Polyacetylene compounds, thujone CC Camphor, eucalyptol
2011	Artemisia herba-alba Asso.	AP Thujone CC Camphor, eucalyptol

CoE No.	Latin name	Active principles (AP)/ Other chemical components (CC)
78	Aspidosperma quebracho-blanco Schlechtend.	Not known
86	Berberis vulgaris L.	Not known
91	Boronia megastigma Nees ex Bartl.	Not known
93	Boswellia sacra Fleckiger	CC Eucalyptol
236	Bursera ssp.	Not known
103 b	Cananga odorata Hook. fil. et Thomson f. macrophylla	Not known
109	Carica papaya L.	Not known
112	Carum carvi L.	Not known
3002	Castor fiber L.	Not known
183	Centaurium erythraea Rafn.	AP Xanthones
2027	Cinchona officinalis L.	CC Quinine
128	Cinchona pubescens Vahl	CC Quinine
134	Cistus ladanifer L.	AP Thujone CC Eucalyptol
141	Citrus aurantiifolia (Christm.) Swingle	AP Furocoumarins CC Eucalyptol
136 a	Citrus aurantium L. ssp. Aurantium L., rind	AP Furocoumarins
136 b	Citrus aurantium L. ssp. Aurantium L., flower	Not known
136 c	Citrus aurantium L. ssp. Aurantium L., leaf and twig	Not known
137	Citrus auriantum L. ssp. Bergamia (Risso&Poit.)	AP Furocoumarins
138	Citrus auriantum L. var. myrtifolia Ker-Gawl.	Not known
2032	Citrus japonica Thum.	Not known
139a	Citrus limon (L.) Burm. rind	AP Furocoumarins
139b	Citrus limon (L.) Burm. leaf and twig	Not known
2035	Citrus medica L. var. medica	Not known
142	Citrus reticulata Blanco	Not known
2039	Citrus reticulata Blanco var. deliciosa H.H.Hu	Not known
2031	Citrus reticulata Blanco var. unshiu (Marco.) H.H.Hu	Not known
143	Citrus sinensis (L.) Pers.	Not known
140	Citrus x paradisi Macfad.	Not known
147	Cocos nucifera L.	Not known
149	Cola acuminata (P.Beauv.) Schott&Endl.	CC Caffeine
2041	Cola nitida (Vent.) Schott&Endl.	CC Caffeine
155	Corylus avellana L.	Not known
163	Curcuma longa L.	CC Eucalyptol

CoE No.	Latin name	Active principles (AP)/ Other chemical components (CC)
38	Cymbopogon citratus (DC.) Stapf	AP Thujone CC Eucalyptol
2045	Cymbopogon flexuosus (Nees ex Steud.) W. Wats.	AP Methyleugenol
40	Cymbopogon martinii (Roxb.) W.Wats. var. martinii	AP Estragole
39	Cymbopogon nardus (L.) W.Wats.	AP Methyleugenol
2046A	Cymbopogon winterianus Jowitt	Not known
185	Eucalyptus globulus Labill.	CC Eucalyptol
194	Evernia prunastri (L.) Ach.	AP Thujone CC Camphor
196	Ferula assa-foetida L.	Not known
197	Ferula gummosa Boiss.	Not known
2032	Fortunella japonica (Thunb.) Swingle	Not known
77	Galium odoratum (L.) Scop.	AP Coumarin
210	Gardenia jasminoides Ellis	AP Geniposide
214	Gentiana lutea L.	AP Xanthones
230	Hierochloe odorata L.	AP Coumarin
232	Hordeum vulgare L.	Not known
234	Hypericum perforatum L.	AP Hypericine, xanthones
238	Illicium verum Hook.	AP Estragole, safrole
245	Jasminum grandiflorum L.	Not known
246	Jasminum officinale L.	Not known
255	Laurus nobilis L.	AP Methyleugenol CC Eucalyptol
270	Mangifera indica L.	Not known
272	Marsdenia cundurango Rchb. f.	AP Coumarin
2076	Murraya koenigii (L.) Spreng.	Not known
294	Musa L. species, Musa sapientum L.	Not known
309	Olea europaea L.	Not known
345	Piper cubeba L.	Not known
352	Plantago lanceolata L.	Not known
361	Populus nigra L.	Not known
388	Quercus alba L.	Not known
2100	Saccharum officinarum L.	Not known
417	Sambucus nigra L.	AP Hydrocyanic acid
420	Santalum album L.	Not known
427	Schinus molle L.	CC Carvacrol
440	Swertia chirata Buch.-Ham. ex Wall.	AP Xanthones
2112	Triticum aestivum L. emend. Fiori et Paol.	Not known

CoE No.	Latin name	Active principles (AP)/ Other chemical components (CC)
468	Urtica dioica L.	Not known
470	Vaccinium macrocarpon Ait.	Not known
469	Vaccinium myrtillus L.	Not known
471	Vaccinium uliginosum L.	Not known
473	Valeriana officinalis L.	Not known
474	Vanilla planifolia G.Jacks.	Not known
479	Vetiveria zizanoides (L.) Nash	Not known
482	Viola odorata L.	Not known
483	Viola tricolor L.	Not known
3006	Viverra zibetha Schreber	Not known
488	Zea mays L.	CC Carvacrol

Abelmoschus moschatus Moench

CE No.	1
Steinmetz No.	1
FEMA No.	Ambrette seed absolute: 2050; Ambrette seed oil: 2051; Ambrette tincture: 2052
Order	Malvales
Family	Malvaceae
Name	E Ambrette, musk mallow
	F Ambrette, abelmose, herbe musquée
	D Ambrette, Abelmoschus
	I Ambretta
	SP Abelmosco
Synonyms	Hibiscus abelmochus L.

Parts used Seed

Important constituents Essential oil of seed
Esters: mainly (2E, 6E)-farnesyl acetate 39-59.1%, (2Z, 6E)-farnesyl acetate 3.8-5,8%, (2E, 6E)-farnesyl propionate 1.1%, (2E, 6E)-farnesyl valerate 0.3%, decyl acetate 1.78-5.6%, hexyl propionate 0.2%, octyl butyrate 0.2%, dodecyl acetate 4.0%, dodecenyl acetate 0.2-2.13%, tetradecenyl acetate 1.6%, methyl linoleate 0.3%, ethyl linoleate 0.2%
Olides: mainly ambrettolide = (Z)-hexadec-7-en-16-olide 7.8-8.8%, (Z)-tetradec-5-en-14-olide 1.8%, (Z)-octadec-9-en-18-olide 1.0%
Alcohols: decanol 0.6%, octanol 0.1%, dodecanol 0.3%, (E) nerolidol 0.4-1.2%, (2E, 6E)-farnesol 3.5%, (2Z, 6E)-farnesol 0.3%
Acids: linoleic acid 0.5%, hexadecanoic acid 1.4%, (E)-octadec-9-enoic acid 0.9%, octadecanoic acid 0.1%
Carbonyle compounds: (2E, 4E)-decadienal 0.8%, (2E, 4Z)-decadienal 0.4%, decan-2-one 0.2%, geranylacetone 0.3% (1,2)
Concrete of seed
Esters: mainly (2E, 6E)-farnesyl acetate 30.9%, (2Z, 6E)-farnesyl acetate 2.25%, (2E, 6E)-farnesyl myristate 0.54%, (2E, 6E)-farnesyl stearate 0.3%, (2E, 6E)-farnesyl palmitate 2.54%, (2E, 6E)- farnesyl oleate 4.0%, (2E, 6E)-farnesyl linoleate 0.32%, decyl acetate 1.21%, dodecyl acetate 1.21%
Olide: ambrettolide 5.1%
Alcohols: dodecanol 0.13%, (E) nerolidol 0.22%, (2E, 6E)-farnesol 2.16%
Acids: octanoic acid 0.07%, nonanoic acid 0.1%, dodecanoic acid 0.18%, tetradecanoic acid 1.08%, hexadecanoic acid 19.08%, (E)-octadec-9-enoic acid 0.21% (3)

	Oleoresin of seed Esters: mainly (2E, 6E)-farnesyl acetate 56.6%, (2Z, 6E)-farnesyl acetate 3.7%, (2E, 6E)-farnesyl propionate 1.6%, (2E, 6E)-farnesyl valerate 0.4%, decyl acetate 1.8%, octyl butyrate 0.1%, dodecenyl acetate 0.1%, decyl propionate 0.1%, tetradecenyl acetate 0.4%, methyl linoleate 0.2%, ethyl linoleate 0.2% Olides: mainly ambrettolide = (Z)-hexadec-7-en-16-olide 9.1%, (Z)-tetradec-5-en-14-olide 0.4%, (Z)-octadec-9-en-18-olide 5.0% Acids: linoleic acid 2.0%, hexadecanoic acid 2.1%, (E)-octadec-9-enoic acid 6.9%, octadecanoic acid 1.2% (1) Volatile fraction of oleoresin Esters: mainly (2E, 6E)-farnesyl acetate 57.6%, (2Z, 6E)-farnesyl acetate 3.2%, (2E, 6E)-farnesyl propionate 1.1%, farnesyl valerate 0.1%, decyl acetate 4.2%, hexyl propionate 0.1%, octyl butyrate 0.1%, dodecyl acetate 4.1%, dodecenyl acetate 0.8%, tetradecenyl acetate 1.6%, methyl linoleate 0.2%, ethyl linoleate 0.1% Olides: mainly ambrettolide = (Z)-hexadec-7-en-16-olide 9.9%, (Z)-tetradec-5-en-14-olide 0.7%, (Z)-octadec-9-en-18-olide 2.1% Alcohols: decanol 0.2%, octanol 0.1%, dodecanol 0.2%, (E) nerolidol 0.1%, (2E, 6E)-farnesol 3.1%, (2Z, 6E)-farnesol 0.2% Acids: linoleic acid 0.3%, hexadecanoic acid 1.0%, (E)-octadec-9-enoic acid 1.1%, octadecanoic acid 0.2% Carbonyl compounds: (2E, 4E)-decadienal 0.5%, (2E, 4Z)-decadienal 0.2%, decan-2-one 0.1%, geranylacetone 0.1% (1)
Active principles	Not known
Other chemical components	Not known
Products in which used	Baked goods, frozen dairy, soft candy, gelatin, puddings, non-alcoholic and alcoholic beverages, chewing gum (4)
Level of use	Ambrette absolute oil: baked goods 2.19 ppm; frozen dairy 1.94 ppm; soft candy 2.48 ppm; gelatin, puddings 2.84 ppm; non-alcoholic beverages 1.51 ppm; alcoholic beverages 7.0 ppm (4) Ambrette seed oil: baked goods 0.63 ppm; frozen dairy 1.07 ppm; soft candy 1.4 ppm; gelatin, puddings 0.5 ppm; non-alcoholic beverages 1.2 ppm; alcoholic beverages 3.1 ppm; chewing gum 3.75 ppm (4) Ambrette tincture: baked goods 45 ppm; frozen dairy 26.4 ppm; soft candy 40.8 ppm; gelatin, puddings 45 ppm; non-alcoholic beverages 12.4 ppm; alcoholic beverages 26.5 ppm (4)
Preparation	Essential oils, concrete, tincture
Main toxicological data	No relevant data found
Data needed	No data required
Specific observations	None
Classification and limits	**Ambrette and preparations: category 2**

National/int. evaluation	Ambrette seed	CFR 182.10
	Ambrette absolute oil	CFR 182.20
	Ambrette seed oil	CFR 182.20
	Ambrette tincture	CFR 182.20

Main references   (1) Flav. Fragr. J., *7*, p.65, (1992)
(2) Parfum Cosmet. Arômes, *84*, p.77, (1988)
(3) Rivista Ital.EPPOS, *60*, p.606, (1978)
(4) Fenaroli, *1*, p.29, (1995)

Data bases used   Chemical Abstracts (1965-97)
Keywords: Ambrette, Abelmoschus moschatus, Hibiscus abelmoschus

Acer saccharum Marsh

CE No.	11
Steinmetz No.	12
FEMA No.	
Order	Sapindales
Family	Aceraceae
Name	E Black sugar maple, sugar maple tree
	F Erable noir à sucre
	D Zuckerahorn
	I Acero nero, acero zuccherino
	SP Arce de azucar
Synonyms	Acer saccharum L.

Parts used	Sap (from wood)
Important constituents	Sucrose, glucose, fructose (1) and volatile components: alcohols, aldehydes and acetals, ketones, furans, pyrazines (2,3)
Active principles	Not known
Other chemical components	Not known
Products in which used	Maple syrup is a foodstuff consumed as such
Level of use	Not known
Preparation	Syrup prepared by heating of sap
Main toxicological data	No relevant data found
Data needed	No data required
Specific observations	None
Classification and limits	**Maple syrup: category 1**
National/int. evaluation	None
Main references	(1) Arctander. Perfume&Flavour Materials of Natural origin, p.398 (1960)
	(2) J. Agr. Food Chem. 38, p.1242 (1990)
	(3) Proceedings 5th International Conference Porto Karras, Chalkidiki (Greece) 1-3 July 1987, p.241.
Data bases used	Chemical Abstracts, Pascal (1965-91)

Achillea millefolium L.

CE No.	12
Steinmetz No.	13
FEMA No.	
Order	Campanulales
Family	Compositae
Name	E Milfoil, yarrow
	F Mille-feuille
	D Gemein Schafgarbe
	I Achillea millefoglie
Synonyms	A. cuspidata DC.; A. lanulosa Nutt.; Santolina millefolium Bn.
Parts used	Herb, flowers
Important constituents	Fresh plant oil: terpenes [α-pinene 0.63%, camphene 0.79%, sabinene 10.20%, β-pinene 2.86%, myrcene 3.35%, α-phellandrene 0.81%, α-terpinene 0.09%, p-cymene 0.66%, limonene 1.22%, γ-terpinene 0.83%, β-caryophyllene 3.96%, α-humulene 2.77%, germacrene D 25.40%, chamazulene 25.50%]; ketones [α-thujone 0.28%, β-thujone 1.60%, camphor 2.93%]; 1,8-cineole 2.24%, bornyl acetate 0.85% (1)
	Dried plant oil: terpenes [α-pinene 1.66%, camphene 2.20%, sabinene 8.31%, β-pinene 5.69%, myrcene 6.00%, α-phellandrene 0.36%, α-terpinene 0.09%, p-cymene 1.05%, limonene 0.95%, γ-terpinene 1.42%, β-caryophyllene 3.34%, α-humulene 0.60%, germacrene D 4.37%, chamazulene 34.60%]; ketones [α-thujone 0.40%, β-thujone 3.21%, camphor 4.43%]; 1,8-cineole 4.54%, bornyl acetate 3.12% (1)
	Flower oil: terpenes [α-pinene 1.65-9.4%, camphene 1.76%, sabinene 4.49-12.3%, β-pinene 4.49-7.1%, myrcene 3.75%, α-terpinene 0.19-1.3%, p-cymene 0.73-3.7%, limonene 0.07-1.7%, γ-terpinene 2.83-3.7%, β-caryophyllene 2.84%, α-humulene 0.11%, germacrene D 21.60%, chamazulene 26.70%, camphene 6%, allo-ocimene 1.4%]; ketones [α-thujone 1.02%, β-thujone 0.59%, camphor 8.01%, isoartemisia ketone 8.6%]; 1,8-cineole 3.70-9.6%, bornyl acetate 2.50%, linalyl acetate, linaleol, geraniol, camphor 17.8%, trerpinen-4-ol 4.3% (1)
	Leaf oil: terpenes [a-pinene 3.10%, camphene 5.14%, sabinene 5.95%, β-pinene 13.90%, myrcene 6.30%, α-terpinene 0.39%, p-cymene 0.73%, limonene 1.25%, γ-terpinene 2.30%, β-caryophyllene 2.84%, α-humulene 0.18%, germacrene D 10.80%, chamazulene 11.10%]; ketones [α-thujone 0.50%, β-thujone 0.25%, camphor 16.80%]; 1,8-cineole 6.09%, bornyl acetate 3.02% (1)

	Stem: achillin, β-sitosterol, stigmasterol, campesterol, α-amimrin, taraxasterol, pseudotaraxasterol
Active principles	Thujone
Other chemical components	Camphor, eucalyptol
Products in which used	Wine and liqueur industries in the formulation of bitters and vermouths
Level of use	Yarrow herb: non-alcoholic beverages 29 ppm, alcoholic beverages 5-40 ppm
Preparation	Essential oil, fluid extract, soft extract, tincture (10% in 65-70% ethanol), infusion (5%), decoction (8%)
Main toxicological data	Thujone: cf datasheet CE No.60, Artemisia abrotanum L. Achillin orally possess hypothermic activity in the rat
Data needed	No data required
Main references	(1) Perf. Flav., 22(3), p.68, (1997) (2) Haggag, M.Y. et al. Planta Med., 27, 361-366, 1975 (3) Tewari, J.P. et al. Ind. J. Med. sci. 28, 331-336, 1974 (4) Kozlowski J. & J. Lutomski, Planta Med., 17, 226-229, 1969 (5) Chandler R.F. et al. J. Pharm. Sci., 71, 690-693, 1982 (6) Falk, A.J. et al. Lloydia, 37, 598-602, 1974
Specific observations	Use of varieties that contain low amounts of thujones. Very high amounts of thujones in some wild Canadian Achillea millefolium L. harvested in Quebec
Classification and limits	**Herb, flowers, essential oil and other preparations: category 4 (with limits on camphor, eucalyptol and thujone)**
National/int. evaluation	FDA §121.1163. In beverages only; finished beverage must be thujone-free *Evaluation of thujone:* No ADI (JECFA, 25th Session, 1981). CEE Directive 88-388: limits: foods: 0.5 mg/kg; beverages: 0,5 mg/kg; alcoholic beverages <25% alcohol: 5 mg/kg; alcoholic beverages >25% alcohol: 10 mg/kg; sage stuffings and food containing sage: 25 mg/kg; bitters: 35 mg/kg Council of Europe: thujone = active principles Provisional limits set up at the 39th Meeting: food: 0.1 mg/kg; beverage: 0.1 mg/kg; alcoholic beverages <25% alcohol: 2 mg/kg; alcoholic beverages >25% alcohol 10 mg/kg; sage stuffings and food containing sage: 10 mg/kg; vinegar with herbs: 5 mg/kg; sweets: 50 mg/kg
Data bases used	Chemical Abstracts – 1997 Toxline Medline Embase Biosis Cancer line FSTA

Aframomum melegueta K.Schum.

CE No.	15
Steinmetz No.	38
FEMA No.	Seeds: 2529
Order	Scitamineae
Family	Zingiberaceae
Name	E Grains of paradise, guinea grain, guinea pepper, melegueta, malagueta or melagueta pepper, alligator pepper F Graines de paradis, maniguette D Paradieskörnerpflanze, Melegueta-Pfeffer I Grana paradisi SP Melegueta
Synonyms	-
Parts used	Seeds (1,2)
Important constituents	6- and 8-gingerols, and 6- and 8-paradols, which make up to 3% of the total weight of the dry seed and occur in approximately equal proportions; 6-shogaol (3). Seed essential oil: eugenol, other phenols (4)
Active principles	Not known
Other chemical components	Not known
Products in which used	Pepper (1), vinegar (1), condiments (1,2), spirits/liqueurs (1,2). Also used as a pepper substitute and used to season food by West Africans (1)
Level of use	Seeds: Non-alcoholic beverages 100 ppm (2) Meat products 3 ppm (2)
Preparation	Dried seed (1,2). From seeds: tincture (20% in 75% ethanol), distillate (80% ethanol), essential oil (2)
Main toxicological data	No toxicity data found for extracts of seeds. Individual components 6-gingerol and 6-shogaol both positive in Ames Test (5,6)
Data needed	Quantitative data on chemical components, gingerol and shogaol, of seed extract and, if necessary, 28-day oral study and mutagenicity studies on seed extract
Specific observations	Major components common to other acceptable flavouring source materials eg. Zingiber officinale
Classification and limits	Seeds: category 5

National/int. evaluation	UK FACC (1976) Appendix 2, GRAS (I), seeds FEMA No.. 2529
Main references	(1) Tanaka's Cyclopedia of Edible Plants of the World. Tokyo: Keigaku (1976) (2) Fenaroli (1995) (3) J. Chromatogr. 67, 29-35. (1972) (4) MAFF (1995) (5) Mut. Res. 122 2, 87-94 (1983) (6) Cancer Letters, 36, 221-233 (1987)
Data bases used	Chemical Abstracts (1967-90) FSTA(1969-90) Toxline, Toxlit, Toxnet(1981-90) Toxlit65(1965-90) Embase(1974-90) Biosis (1973-90) *Keywords:* Aframomum (w) melegueta, Grains (w) paradise

Agropyron repens (L.) Beauv.

CE No.	19
Steinmetz No.	42
FEMA No.	Dog grass extract: 2403
Order	Graminales
Family	Gramineae
Name	E Dog grass, couch grass, quick grass
	F Petit chiendent
	D Kriechende Quecke, gemeine Quecke
	I Gramigna
	SP Grama el norte
Synonyms	-

Parts used	Rhizomes (1)
Important constituents	Steam distillate: about 44% carvacrol (10.8%), trans-anethole (6.8%), carvone (5.5%), thymol (4.3%), menthol (3.5%), menthone (1.4%), p-cymene (1.1%); furthermore: 2-hexyl-3-methyl-maleic anhydride (1.8%), farnesylacetone, hexahydro farnesylacetone (1.1%), pelargonic acid (1.8%), myristic acid (1.3%), pentadecane acid (2.6%), palmitic acid (23.5%), oleic acid (1.6%), linoleic acid (2.5%) (2), in addition: androstenone (37-42 ng/g dry wt leaf & stem), progesterone (15 ng/g dry wt stem)
	Total phenolic glucosides: [92 µg/g dry wt of roots: 5-hydroxyindole-3-acetic acid (3900 µg/g dry wt; growth hormone), 5-hydroxytryptophan (420 µg/g dry wt; growth inhibitor)] (3, 4, 5)
	Acetylenic compounds: (agropyren, capillen) earlier thought to be main components in the essential oil to which a mild antibiotic action is attributed (1,6) not found in a recent gc-ms investigation (5)
Active principles	Not known
Other chemical components	Carvacrol
Products in which used	Used as a herbal tea (6). Dog grass extracts, alone or in combination with other flavour ingredients. Used to a limited extent in flavouring baked pastry, ice-creams, non-alcoholic beverages, candies, gelatin and puddings (1)
Level of use	Dog grass extracts in non-alcoholic beverages 9.58 ppm, frozen dairy 5.00 ppm, soft candy 5.0 ppm, baked goods 32.42 ppm, gelatin, puddings 0.90 ppm (1)

Preparation	Fluid extract, soft aqueous extract, dried aqueous extract, tincture, infusion (1)
Main toxicological data	No toxicity data found for extracts of rhizomes Acetylenic compounds (agropyren and capillen): active compounds
Data needed	A 28-day study needed and mutagenicity studies on the relevant extract. Further data needed on toxicology, content and occurrence of the acetylenic compounds, agropyren and capillen
Specific observations	None
Classification and limits	**Rhizomes and preparations: category 5 (with limits on carvacrol)**
National/int. evaluation	Dog grass extract: CFR 182.20, 582.20
Main references	(1) Fenaroli (1995) (2) Planta Medica 55: 399-400 (1989) (3) Can. J. Bot. 67 (2): 288-296 (1989) (4) J. Agric. Food Chem. 37: 1143-1149 (1989) (5) Planta Medica 55: 399 (1989) (6) Wichtl M., Teedrogen, Stuttgart (1989)
Data bases used	Chemical Abstracts (1967-93) Medline (1966-97) Embase (1980-97) Biological Abstracts (1989-97) Toxline (1965-93) *Keywords:* Agropyron repens, Couch-grass, Agropyren

Aloysia triphylla (L'Herit.) Britt.

CE No.	264
Steinmetz No.	662
FEMA No.	-
Order	Tubiflorae
Family	Verbenaceae
Name	E Vervain
	F Verveine odorante
	D Zitronenstrauch, Zitronenkraut, echte Verbene
	I Cedrina
	SP Hierba luisa
Synonyms	Lippia citriodora (Ort. ex Berg.) H.B.K., Lippia triphylla (L'Herit.) Kuntze, Verbena triphylla L'Her
Parts used	Herb, leaves
Important constituents	5-hydroxy-6,7,4'-trimethoxyflavone (salvigenine), 5,3'-dihydroxy 6,7,4'-trimethoxyflavone (eupatorin), 5,7,3'4'-tetrahydrocy-6-methoxyflavone (eupafolin), 6-hydroxyluteolin, luteolin, luteolin-7-o-b-glucoside, 4',5,7-trihydroxy-6-methoxyflavone (hispidulin), 4',5-dihydroxy-6,7-dimethoxyflavone (cirsimaritin), diosmetin, chrysoeriol, apigenin, 5,7-dihydroxy-6,4'-dimethoxyflavone (pectolin-arigenin), 5,3'4'-trihydroxy-6,7-dimethoxyflavone (cirsiliol), caryophyllane-2-6-b-oxide, caryophyllene, caryophyllene epoxide, iso-caryophyllene, iso-caryophyllene epoxide, hex-4-en-1-al-2-5-dimethyl-2-4-inyl, kobusone, cis-limonene epoxide, trans-limonene epoxide, nerol epoxide, perillen, photocitral A, photocitral B, epi-photocitral A, trans-photocitral, photonerol A, epophotonerol A, photonerol B, cis-rose oxide, rosefuran, citral (35%), cineol, geraniol, citronal (1,2,3,4,5,6)
Active principles	Not known
Other chemical components	Not known
Products in which used	The leaves are used for flavouring beverages, dessert, fruit salads and jellies, for seasoning food and in the preparation of tisanes and in sachets
Level of use	No information available
Preparation	None
Main toxicological data	Apigenin and diosmetin are not mutagenic in the Salmonella/mammalian microsome test (7). Administration of citral to rats for 13 weeks at levels of 1000, 2500, or 10 000 ppm in the diet caused no

	macroscopic changes (8). In citral treated rats hepatomegaly was accompanied by an altered distribution of lipid and glycogen in the liver and peroxisome proliferation occurred in a manner reminiscent of that of some hypolopodaemic compounds (9)
Data needed	Level of use and 28-day oral study on herb and leaves
Specific observations	None
Classification and limits	**Herb: category 5** **Leaves: category 5**
National/int. evaluation	None
Main references	(1) Helv.Chim.Acta 59:1797 (1976) (2) Helv.Chim.Acta 59:1803 (1976) (3) Helv.Chim.Acta 59:1802 (1976) (4) Phytochemistry 367 (1971) (5) Planta Med. 54:465 (1988) (6) The wealth of India 6:142 (1962) (7) Mut.Res. 75:243 (1980) (8) Food Cosmet.Toxic. 5:141 (1967) (9) Food Chem.Toxic. 25(7):505 (1987)
Data bases used	Not specified

Amyris balsamifera L.

CE No.	33
Steinmetz No.	-
FEMA No.	-
Order	Rutales
Family	Rutaceae
Name	E West Indian sandalwood
	F Bois de santal des Indes occidentales
	D Balsambaum, Sandelholzbaum
	I Sandalo delle Indie
	SP Amiris, extracto
Synonyms	-
Parts used	Wood
Important constituents	Essential oil of wood:
	Alcohols: valerianol 21.5%; elemol 9.1%: cadinol 50.1%; β-eudesmol 7.9%; γ-eudesmol 6.6%; 10-epi-γ-eudesmol 9.7%; α-eudesmol 4.8%; 7-epi-α-eudesmol 10.7%; dridemol 1.1%
	Terpenic hydrocarbons: cadinene 10.7%; curcumene 1.5%; β-sesquiphellndrene 4.7%; β-bisabolene 0.8%; α-zingiberene 2.4%; seline 3,7(11)-diene 2.5%
	Carbonyl and lactonic compounds: α-agarofuran 0.5%; bisabolone 0.9% (1,2,3)
Active principles	Not known
Other chemical components	Not known
Products in which used	Liqueurs and beverages
Level of use	Liqueurs 0.150 g (wood)/l; beverages 0.100 g(wood)/l
Preparation	Essential oil
Main toxicological data	No data available
Data needed	28-day oral study and study on mutagenicity
Specific observations	Data on wood and essential oil are missing
Classification and limits	Wood: category 5
	Essential oil: category 5
National/int. evaluation	None

Main references (1) Phytochemistry 16, p.773 (1977)
(2) Phytochemistry 28, p.1909 (1989)
(3) Perfumer&Flavorist 15, p.78 (1990)

Data bases used Chemical Abstracts (1972-91)

Angelica archangelica L.

CE No.	56
Steinmetz No.	120
FEMA No.	-
Order	Umbelliflorae
Family	Umbelliferae
Name	E Angelica
	F Angélique
	D Echte Engelwurz, Erzengelwurz, Brustwurz, Angelikawurz
	I Angelica
	SP Angelica
Synonyms	Archangelica officinalis (Moench) Hoffm.

Parts used	Roots, leaves, fruit, stem
Principal components	Roots: polysaccarides (mainly sucrose), tannins, organic acids (angelic = methylcrotonic acid, caffeic acid, chlorogenic acid); flavonoids (archangelenone); essential oil (yield 0.35-1.30%) (1); coumarins and furocoumarins (osthenol 0.37-0.47 µg/g (2); umbelliferone, bergapten, isoimperatorin, xanthotoxin, angelicin, archangelicin)
Essential oil from root: terpenic hydrocarbons (α-pinene 4.4-24%, α-phellandrene 7.5-20%, β-phellandrene 16-24%, β-pinene 0.2-0.5%, sabinene 0.4-4.9%, δ-3-carene 4.5-13%, myrcene 1.6-5.5 %, limonene 13.2%, terpinolene 0.7-2.2%, cis-ocimene 1-1.9%, trans-ocimene 2.4-4.9%, γ-terpinene 0-0.3%, camphene 0.2-1.3 %, β-bisabolene 0.72%, β-caryophyllene 0.4%, β-elemene 0.9%; p.cymene 3.2-11.3%); macrocyclic lactones 0.8-2.4% (tri-, penta- and hepta-decanolides); cis and trans-verbenol 0.1-0.7%, α-phellandren-8-ol 0.4%, β-eudesmol 0.21%, bornyl acetate 0.73%, trans-verbenyl acetate 0.46%, trans-verbenyl isovalerate 0.17%, (3); 2-nitro-1,5 p.menthadiene, cis and trans 6-nitro-1(7)-2-p.menthadiene, trans-1(7)-5-p.menthadiene-2-yl acetate and 7-isopropyl-5-methyl-5-bicyclo (2.2.2) octan-2-one (4)
Alcoholic extract of angelica root: (analysis of volatile only): same qualitative composition as angelica root oil, with some additional esters and terpenic derivatives (5)
Angelica seed oil: (qualitative composition): α-pinene, camphene, β-pinene, myrcene, β-phellandrene, ocimene (no isomer given), limonene, γ-terpinene, caryophyllene, p.cymene, 1,8-cineole, borneol, carvone (ref 3). The seed oil is said to contain 0.5% imperatorin, 0.1% bergapten, 0.02% xanthotoxol, 0.04% umbelliprenin, osthole, osthenole, and angelicin (6)

	Main components in the essential oils obtained from various parts of Angelica archangelica of Romanian origin: Root oil: α-pinene 27%, β-pinene 57.1%; Stem oil: caryophyllene 49.2%; Leaf oil: α-pinene 36.9%, β-pinene 23.9%, β-phellandrene 33.8%; Flower oil: β-phellandrene 68.3%; Fruit oil: β-phellandrene 87.4% (3)
Active principles	Coumarin, furocoumarins
Other chemical components	Not known
Products in which used	Baked goods, frozen dairy, candy, gelatin, puddings, alcoholic and non-alcoholic beverages
Level of use	(7)

	Angelica root extract (ppm)	Angelica root oil (ppm)	Angelica seed extract (ppm)	Angelica seed oil (ppm)	Angelica stem oil (ppm)
Baked goods	56.24	38.68	63.50	68.73	129.60
Frozen dairy	60.82	26.82	60.00	59.08	88.42
Soft candy	60.44	37.42	26.80	70.26	132.20
Gelatin, puddings	91.00	35.81	130.00	54.70	17.50
Nonalc. bev.	54.76	13.31	1200.00	33.00	13.29
Alc. beverages	15.79	41.29	2000.00	75.61	77.00

Preparation	Root fluid extract; tincture (20% of root extract); root essential oil; seed extract in 60% ethanol; seed essential oil; stem essential oil
Main toxicological data	Angelica root oil: Low acute oral toxicity in rats and mice. Low acute dermal toxicity in rabbits. Administered orally to rats daily for 8 weeks at doses of 0.5 – 1.0 – 2.0 or 3.0 g/kg bw, it was concluded that the tolerated dose was 1.5 g/kg bw. Daily doses of 2.0 or 3.0 g/kg bw caused decreased activity and weight loss, and deaths were associated with severe liver and kidney damage. It was pointed out that 1.5 g/kg/day was not truly tolerated for 8 weeks, as rats receiving the 0.5 or 1.0 g/kg daily doses, weighed less than the control animals. Angelica root oil applied undiluted to the backs of hairless mice or to intact or abraded rabbit skin for 24 hr under occlusion was not irritating. A maximisation test was carried out on 24 volunteers. The material was tested at a concentration of 1% in petrolatum and produced no sensitisation reactions. Phototoxic effects were reported for angelica root oil tested on hairless mice and swine. Various concentrations of angelica root oil in methanol were also tested for phototoxicity in mice: applications of 20 µl/ 5 cm2 of skin being exposed to simulated sunlight for 1 hr. Positive reactions were obtained with concentrations of 3.125 – 6.25 – 12.5 – 25 – 50 and 100%, while 1.56% evoked a doubtful reaction and 0.78% showed no phototoxic effect (8)
Data needed	No data required
Specific observations	None

Classification and limits	Stem: category 1 Other parts and preparations: category 4 (with limits on coumarin and furocoumarins)
National/int. evaluation	None
Main references	(1) Fitoterapia, *58*, p.129 (1987) (2) Deutsche Apotheker Ztg, *121*, p.386 (1981) (3) Perfumer&Flavorist, Dec 1976/Jan 1977, *1*, p.31; Jun/July 1981, *6*, p.46 (4) Helv. Chim. Acta, *62*, p.2061 (1979) (5) Acta Chem. Scand., *29B*, p.757 (1975) (6) Duke, p.43 (1989) (7) Fenaroli *1*, p.31 (1995) (8) Food Cosmet. Toxicol., *13 Suppl.*, p.713 (1975
Data bases used	Chemical Abstracts – June 1996

Annona cherimola Mill.

CE No.	2008
Steinmetz No.	-
FEMA No.	-
Order	Magnoliales
Family	Annonaceae
Name	E Custard apple, Cherimoya
	F Chérimole, Chéri molier
	D Cherimoya
	I Pomocarmella, Annona
	SP Chirimoya
Synonyms	-
Parts used	Fruit
Important constituents	Fruit: water ~ 82%, total sugars (sucrose and fructose ~ 16%); fatty acids; amino-acids mainly citrulline and γ-aminobutyric acid; vitamine C, carotenoids; minerals; cyanogenic glucosides (trace), pectin + volatile components (1, 2)
	Volatile components of pulp (~7-10 mg/kg of fruit):
	Terpenic hydrocarbons: mainly *p*-cymene, α-thujene, camphene, α- and γ-terpinene, limonene
	Esters: mainly butyl butanoate, hexyl butanoate, 3-methylbutyl butanoate
	Alcohols: mainly 3-methyl-1-butanol, 1-hexanol, 1-butanol, linalool, 3-methyl-2-butanol, 1-pentanol, 1-penten-3-ol, (*E*)-3-penten-1-ol, (*Z*)-2-penten-1-ol, (*Z*)-3-hexen-1-ol, 2-pentanol, 2-methyl-1-propanol, 2-methyl-3-buten-2-ol, (*E*)-2-hexen-1-ol, 2-ethyl-1-hexanol, benzyl alcohol
	Oxygenated terpenic compounds: mainly a-terpineol, chrysanthenone, furanoid and pyranoid linalool oxides, camphor, verbenone, myrtenal, myrtenol, carvone, eucalyptol (3)
Active principles	Hydrocyanic acid
Other chemical components	Camphor and eucalyptol
Products in which used	Pulp consumed as such or used to prepare fresh drinks
Level of use	Not found
Preparation	Juice
Main toxicological data	No relevant data found
Data needed	No data required

Specific observations	Fruit is used as a foodstuff
Classification and limits	**Fruit: category 3 (with limits on camphor, eucalyptol and hydrocyanic acid)**
National/int. evaluation	None
Main references	(1) Phytochemistry, *21*, p.2783, (1982) (2) Z. Lebensm. Unters. Forsch., *173*, p.47, (1981) (3) J. Agric. Food Chem., *32*, p.383, (1984)
Data bases used	Chemical Abstracts (1965-97) *Keywords:* Cherimoya, Custard apple, Annona cherimola

Annona squamosa L.

CE No.	46
Steinmetz No.	101
FEMA No.	-
Order	Magnoliales
Family	Annonaceae
Name	E Sweetsop, sugar apple, custard apple
	F Pomme-cannelle, annone écailleuse
	D Rahmapfel, Zimtapfel, Süßsack, Zuckerapfel
	I Annona scaliosa
	SP Anona
Synonyms	-
Parts used	Fruits, seeds
Important constituents	Fruit: carbohydrates, sucrose and fructose; fatty acids; amino-acids mainly citrulline and γ-aminobutyric acid, ornithine, arginine, asparagine, histidine, alanine, tyrosine and lysine; vitamins (vitamine C, carotenoids); cyanogenic glucosides ± volatile components (1)
	Volatile components of pulp (~7-10 mg/kg of fruit):
	Terpenic hydrocarbons: α-pinene 0.8%, β-pinene 1.3%, β-myrcene 0.3%, trans-β-ocimene 0.6%, terpinolene 1.0%, α-copaene 5.2%, β-cubebene 1.7%, δ-cadinene 1.2%
	Carbonyle and lactonic compounds: butanedione 0.3%, pentane 2,3-dione 0.4%, 3-hydroxybutan-2-one 0.9%, γ-butyrolactone 0.1%, γ-hexalactone, verbenone 0.5%; furfural 1.2%
	Terpenic alcohols: T-cadinol 37.0%, T-muurolol 12.4%, eucalyptol 1.4%, borneol 0.3%, terpinen-4-ol 1.3%, α-terpineol 5.1%, farnesol 3.5%, globulol 4.0%, viridiflorol 2.1%, fenchyl alcohol 0.2%
	Ester: bornyl acetate 1.8% (2)
	Seed:
	Oil and fatty acids: oleic, linoleic, myristic, palmitic, stearic, arachidic, palmitoleic, linolenic acids
	Glycerides of one or more hydroxylated fatty acids; albumin and globulin (1)
Active principles	Hydrocyanic acid
Other chemical components	Eucalyptol
Products in which used	Pulp eaten as dessert fruit or can made into drinks and fermented liquor
Level of use	Not found

Preparation	Pulp consumed as such or used to prepare fresh drinks. Seed oil as foodstuff
Main toxicological data	No relevant data found
Data needed	No data required
Specific observations	Fruit is used as a foodstuff
Classification and limits	**Fruit: category 3 (with limits on eucalyptol and hydrocyanic acid)** **Seed oil: category 2**
National/int. evaluation	None
Main references	(1) Phytochemistry, 21, p.2783, (1982) (2) Flav. Frag. J., 8, p.5, (1993)
Data bases used	Chemical Abstracts (1965-97) *Keywords:* Sweetsop, Annona squamosa

Anthoxantum odoratum L.

CE No.	49
Steinmetz No.	
FEMA No.	
Order	Graminales
Family	Gramineae
Name	E Pig grass, perennial sweet vernal grass
	F Floure odorante
	D Ruchgras
	I Paleo odoroso
	SP Grama de olor
Synonyms	
Parts used	Herb, flowers, leaves, roots (1)
Important constituents	Contains coumarin (about 5% of dry matter in freshly dried herb of which about 40% present as free coumarin) (2,3) and o-coumaric acid (trans-2-hydroxycinnamic acid) (4,5). Coumarin derivatives (total of about 0.22% of dry matter in freshly dried herb) such as dicoumarol (less than 0.0001%), 4-hydroxycoumarin, 7-hydroxy-coumarin. Levels of coumarin related compounds are increasing during spoilage in sweet vernal hay to 0.32% of dry matter while coumarin content is decreasing to trace amounts within four weeks (2)
Active principles	Coumarin
Other chemical components	Not known
Products in which used	Used for flavouring purposes in non-alcoholic and alcoholic beverages, ices, candies, baked goods, meat products and desserts (5,6,7; IOFI 1998)
Level of use	Absolute of whole plant: non-alc. beverages 1 ppm, alc. beverages 1 ppm, ices 1 ppm, candies 1 ppm, baked goods 1 ppm, desserts 1 ppm, meat products 1 ppm Essential oil of whole plant: non-alc. beverages 0.26-0.50 ppm, alc. beverages 0.4-2.5 ppm, ices 0.8-1.0 ppm, candies 1.0-2.5 ppm, baked goods 1.2-3.0 ppm, desserts 0.2-1.0 ppm, meat products 0.5 ppm, soups 0.5 ppm, snacks 0.5 ppm (IOFI 1998)
Preparation	Steam-distillate from leaves
Main toxicological data	Report from England of a fatal haemorrhagic diathesis in cattle fed home produced hay containing about 80% sweet vernal grass. A similar syndrome with progressive weakness, stiff gait, mucosal

	pallor, tachycardia, tachypnoe and haematoma with haemorrhages in most organs ending in sudden death was reproduced experimentally in calves fed the hay and was demonstrated to be due to dicoumarol (15 ppm in hay) produced by fungi from o-coumaric acid (8)
Data needed	No data required
Specific observations	None
Classification and limits	**Herb essential oil and absolute: category 4 (with limits on coumarin)**
National/int. evaluation	SCF-opinion on coumarin: general limit for food and beverages recommended at the currently achievable limit of detection for coumarin of 0.5 mg/kg (SCF, Dec. 1996)
Main references	(1) CoE, Flavour. Subst. and Nat. Sources of Flavourings, Strasbourg (1981) (2) J. Sci. Fd. Agric. 15: 733-738 (1964) (3) Agronomy J. 49: 493-497 (1957) (4) Tyler V.E. et al., Pharmacognosy, 9th ed., Lea & Febiger, Philadelphia (1988) (5) Bauer K. et al., Common Fragrance and Flavor Materials, (1990) (6) Cheeke P.R. and Shull L.R., Natural Toxicants in Foods and Poisonous Plants, (1985) (7) pta in der Apotheke 10, (11): 338-342 (1981) (8) The Vet. Record, July 23: 78-84 (1983)
Data bases used	Chemical Abstracts(1982-97) RTECS Toxline (1980-93) Embase (1980-97) Medline (1966-97) Biological Abstracts (1989-97) *Keywords:* Anthoxantum odoratum, Pig grass, Sweet vernal grass.

Anthriscus cerefolium (L.) Hoffm.

CE No.	50
Steinmetz No.	106
FEMA No.	Chervil: 2279
Order	Umbelliflorae
Family	Umbelliferae
Name	E Chervil
	F Cerfeuil
	D Gartenkerbel
	I Cerfolio
	SP Perifollo – cerifolio
Synonyms	Chaerefolium cerefolim (L.) Schinz et Thellung
Parts used	Aerial parts (herb); leaves
Important constituents	Aerial parts: bitter principles; volatile compounds (0.03 %) Essential oil: alkenylbenzenes [estragole (methylchavicol): 75-80 %; 1-allyl-2,4-dimethoxy-benzene 16-22 %; (E)-anethole 0.2 %]; terpenic hydrocarbons [limonene 0.2 %; traces of b-phellandrene; α- + β-pinene; myrcene] (1,2)
Active principles	Estragole
Other chemical components	Not known
Products in which used	Baked goods, frozen dairy, meat products, condiment, relish, soft candy, non-alcoholic beverages
Level of use	Chervil: baked goods 190 ppm; frozen dairy 70 ppm; meat products 1140 ppm; condiment: relish 80 ppm; soft candy 80 ppm; non-alcoholic beverages 120 ppm (3)
Preparation	Essential oil
Main toxicological data	trans-Anethole: ADI 0-2 mg/kg (4)
Data needed	Further data needed on use and use levels of essential oil Toxicity and mutagenicity data on 1-allyl-2,4-dimethoxy-benzene
Specific observations	Leaves are used as a foodstuff
Classification and limits	**Leaves: category 1** **Essential oil: category 5 (with limits on estragole)**
National/int. evaluation	Chervil and Chervil extract: FDA 182.10
Main references	(1) Perf. Flav., *8*, p.72, (June/July 1983)
	(2) J. Essent. Oil Res., *8*, p.305, (1996)

(3) Fenaroli (1995)
(4) JECFA 51st meeting (1998)

Data bases used Chemical Abstracts (1965-June 1996)
Keywords: Chervil, Anthriscus cerefolium.

Artemisia abrotanum L.

CE No.	60
Steinmetz No.	134
FEMA No.	
Order	Campanulales
Family	Asteraceae
Name	E Southern wood
	F Aurône mâle
	D Eberreis, Eberraute
	I Abrotano
Synonyms	
Parts used	Herb
Important constituents	Herb (qualitative composition): coumarinic derivatives [umbelliferone; scopoletine; isofraxidine]; dihydrofurano ketone [davanone]; terpenic alcohol [*trans*-4-thujanol]; ester [sabinyl acetate]; tannins (1); flavonoids [aglycones = luteolin; quercetin 3,3' dimethyl; quercetin-3,7 dimethyl, quercetin-3,7,3' trimethyl; quercetagetin-3,6,4' trimethyl; quercetagetin-3,6,7,4' tetramethyl] (2)
	Essential oil (qualitative composition): terpenic hydrocarbons [a-thujene; α-pinene; β-pinene; camphene; sabinene; fenchene; γ-terpinene; limonene; (E)-ocimene; π-cymene; terpinolene; caryophyllene; humulene; α-copaene; δ-cadinene] (3); oxygenated terpenic compounds [linalool; α-terpineol; 1,8-cineole; piperitol; farnesol or nerolidol (3); α-thujone (4)]; sesquiterpene lactones and dihydrofurano compounds [davanone; nordavanone; *cis*-artemone; davanol (3,4)]; alkenylbenzene [anethole; methyleugenol]; coumarinic derivatives [umbelliferone]; esters [geranyl butyrate; geranyl pentanoate] (3): quantitative compositions: not found
Active principles	Methyleugenol, thujone
Other chemical components	Eucalyptol
Products in which used	Alcoholic and non-alcoholic beverages
Level of use	Alcoholic extract: alcoholic beverages 2 mg/l; non-alcoholic beverages 1 mg/l (IOFI 1995)
Preparation	Alcoholic extract (50 vol. %)
Main toxicological data	Thujone ($\alpha + \beta$): *Acute toxicity:* toxic doses induced epileptic convulsions with general vasodolatation, decrease of blood pressure, lower cardiac rhythm and increase of respiratory amplitude. Short-term studies: 14 weeks gavage studies (isomeric composition of the

material not specified): NOEL = 10 mg/kg/d for male rat and = 5 mg/kg/d for female rat. 13 weeks gavage studies with commercial mixture of α- and β-thujones: convulsive ED50 = 35.5 mg/kg/d for male rat and = 26.3 mg/kg/d for female rat; NOEL = 12.5 mg/kg/d for male rat and NOEL cannot be established for female rat. JECFA: ADI not specified (5).

After oral administration of thujones (as a mixture of α- and β-isomers in the ratio of 9/2) to rabbits, 2 neutral urinary metabolites were obtained: neo-isothujanol (=β-hydroxy-α-thujane) and thujanol (=β-hydroxy-β-thujane) (6,7)

Methyleugenol: Not mutagenic with Ames test and with *Escherichia coli* wp2 reversion test. Positive in the yeast assay, in the *Bacillus subtilis* DNA repair test, and in UDS. In mice treated with methyleugenol, presence of DNA adducts in liver. Hepotocarcinogenic effects occurred in pre-weaning mice treated with methyleugenol or with its 1'-hydroxy metabolite by i.p. injection

Many metabolic pathways: allylic hydroxylation and oxidation of the side chain, *O*-demethylation, hydroxylation of the benzene ring. Ames test and UDS positive with 2,3-epoxy-methyleugenol (8)

JECFA: ADI not specified (5)

Data needed	No data required
Specific observations	None
Classification and limits	Herb: category 4 4 (with limits on eucalyptol, methyleugenol and thujone) Essential oil: category 4 (with limits on eucalyptol, methyleugenol and thujone)
National/int. evaluation	Thujone use restricted as in Annex II of Council Directive 88/388/EEC. Methyleugenol: NTP (National Toxicology Program): Long term and carcinogenis studies on going
Main references	(1) J. Nat. Prod., *45*, 455, (1982) (2) Fitoterapia, *60*, 460, (1989) (3) Z. Lebensm. Unters. Forsch., *179*, 125, (1984) (4) Planta Med., *45*, 55, (1982) (5) JECFA 25th meeting (1981) (6) Xenobiotica, *19*, 843, (1989) (7) Thujone, manuscript CoE (1999) (8) Methyleugenol, manuscript CoE (1999)
Data bases used	Chemical Abstracts Toxline

Artemisia absinthium L.

CE No.	61
Steinmetz No.	135
FEMA No.	Artemisia (wormwood) 3114; Artermisia (wormwood) extract 3115; Artemisia (wormwood) oil 3116
Order	Campanulales
Family	Asteraceae
Name	E Wormwood herb
	F Armoise, Grande absinthe
	D Wermut, Absinth
	I Assenzio
Synonyms	Absinthium officinale Brot. – Artemisia vulgare Lam.

Parts used	Herb
Important constituents	Herb: (qualitative composition): sesquiterpene lactones [absinthin; isoabsinthin; anabsin; anabsinthin; arabsin; arlatin; sesartemin; diasesartemin; episesartemin; absintholide; artabsinolide A, B and C; artenolide; hydroxypelenolide; ketopelenolide A and B]; hydrocarbons [7-ethyl-3,6-dihydro-1,4-dimethyl-azulene; 7-ethyl-5,6-dihydro-1,4-dimethyl-azulene]; ose and osids [inulobiose; spinacetin-3-O-rutinoside; spinacetin-3-O-β-D-glucopyranoside]; acetylenic derivative [5-(1-propynyl)-2-thiophene-propanoic acid, methyl ester] (1,2); flavonoids [aglycones = kaempferol; kaempferol-3-methyl; quercetagetin-3,6,3' trimethyl; quercetagetin-3,6,7 trimethyl; quercetagetin-3,6,7,3' tetramethyl; quercetagetin-3,6,7,3',4' pentamethyl] (3)
	Essential oil [(Z)-epoxy-ocimene chemotype]: terpenic hydrocarbons [a-pinene 0.13-2.04%; sabinene 0.26-6.70%; myrcene trace-5.73%; limonene trace-1.14%; (Z)-epoxy-a-ocimene 25.73-54.38%; (E)-epoxy-α-ocimene 0.81-3.51%; β-caryophyllene 0.60-2.19%; germacrene D 0.60-2.64%; β-curcumene trace-0.89%] (4); oxygenated terpenic compounds [α-thujone trace-0.30%; β-thujone trace-7.78%; camphor 0.19-9.30%; linalool 0.24-2.54%; nerol 0.49-2.17%; sabinol (correct isomer not characterised) trace-2.53%; chrysanthendiol (tentative identification) 0.55-17.25%; spathulenol trace-0.91%; α-bisabolol 1.47-3.38%] (4); . esters [chrysanthenyl acetate trace-10.96%; sabinyl acetate trace-11.52%; neryl acetate trace-0.89%; geranyl acetate 0.39-1.69%; neryl isobutyrate 0.27-3.35%; geranyl isobutyrate trace-0.92%; neryl butyrate 0.28-8.06%; neryl isovalerate trace-0.38%; geranyl butyrate trace-1.21%; geranyl isovalerate trace-0.38%] (4)

Essential oil [sabinyl acetate chemotype]: terpenic hydrocarbons [a-pinene ~ 0.10%; sabinene 0.37-0.83%; myrcene 0.47-2.09%; p-cymene 0.13-0.20%; (Z)-epoxy-α-ocimene trace-0.12%; β-curcumene 0.31-1.65%; β-caryophyllene 0.11-0.82%; germacrene D trace-0.26%] (4); esters [sabinyl acetate 31.47-84.48%; chrysanthenyl acetate trace-5.80%; neryl acetate 0.43-0.79%; geranyl acetate 0.14-4.17%; neryl isobutyrate 0.45-2.59%; geranyl isobutyrate 0.19-0.75%; neryl butyrate 0.10-0.85%; neryl isovalerate 0.22-9.05%; geranyl butyrate 0.10-0.85%; geranyl isovalerate trace-0.77%] (4); oxygenated terpenic compounds [α-thujone 0.12-0.20%; β-thujone 0.58-0.71%; camphor trace-0.31%; linalool 0.60-2.53%; nerol 0.19-3.16%; sabinol (correct isomer not characterised) 0.81-4.18%; chrysanthendiol (tentative identification) trace-0.47%; spathulenol 0.14-1.55%] (4)

Essential oil [chrysanthenyl acetate chemotype]: terpenic hydrocarbons [a-pinene 1.46%; sabinene 0.81%; myrcene 2.90%; p-cymene 0.22%; (Z)-epoxy-α-ocimene trace-0.34%; β-curcumene 0.29%] (4); esters [sabinyl acetate 0.73%; chrysanthenyl acetate 41.54%; neryl acetate 0.62%; geranyl acetate 0.22%; neryl isobutyrate 0.25%; geranyl isobutyrate 0.12%; neryl butyrate 1.10%; neryl isovalerate 0.67%] (4); oxygenated terpenic compounds [α-thujone 1.32%; β-thujone 18.72%; camphor 0.18%; linalool 0.41%; nerol 1.15%; sabinol (correct isomer not characterised) 0.49%; chrysanthendiol (tentative identification) 2.40%; spathulenol 0.12%; α-bisabolol 0.44%] (4)

Essential oil [β-thujone chemotype]: terpenic hydrocarbons [α-pinene 0.16-0.72%; sabinene 1.09-10.63%; myrcene 0.15-1.15%; p-cymene 0.34-0.51%; (Z)-epoxy-a-ocimene 0.05-1.48%; b-caryophyllene 1.47-3.35%; germacrene D 0.31-0.80%] (4); esters [cis-sabinyl acetate 15.09-53.42; trans-sabinyl acetate + lavandulyl acetate 27.78%; geranyl propionate 0.10-1.43%] (5,6,7); oxygenated terpenic compounds [[α-thujone 0.53-2.76%; β-thujone 17.50-59.90%; camphor 0.10-0.16%; linalool 1.15-4.56%; cis-sabinol 0.10-0.47%; caryophyllene oxide 0.80-1.12%] (5,6,7)

Essential oil [β-thujone/epoxy ocimene mixed chemotypes]: terpenic hydrocarbons [α-pinene 0.18-0.34%; sabinene 0.33-6.34%; myrcene 0.26-1.39%; β-ocimene 0.24-0.51%; (Z)-epoxy-α-ocimene 22.39-28.90%; (E)-epoxy-α-ocimene 0.76-1.42%; β-curcumene trace-0.69%; β-caryophyllene 0.22-0.66%; germacrene D 0.40-1.43%] (4); esters [sabinyl acetate 0.27-0.92%; chrysanthenyl acetate 1.49-4.32%; neryl acetate trace-2.12%; geranyl acetate 0.45-1.64%; neryl isobutyrate trace-0.21%; geranyl isobutyrate trace-0.33%; neryl butyrate 0.97-1.30%; neryl isovalerate 0.98-1.19%] (4); oxygenated terpenic compounds [[α-thujone 0.70-1.68%; b-thujone 20.90-40.60%; linalool 0.53-2.15%; nerol 0.35-0.97%; sabinol (correct isomer not characterised) trace-0.19%;

chrysanthendiol (tentative identification) 1.78-6.57%; spathulenol trace-0.69%; α-bisabolol 0.56-2.27%] (4)

Essential oil [*cis*-chrysanthenol chemotype]: terpenic hydrocarbons [[α-pinene 2.94%; sabinene 0.75%; myrcene 10.40%]; oxygenated terpenic compounds [α-thujone 2.55-21.60%; β-thujone 3.75-25.90%; *cis*-chrysanthenol 15.70-69.01%; linalool 2.02%; terpinen-4-ol 1.68%] (8)

Active principles	Thujone
Other chemical components	Camphor
Products in which used	Foods, alcoholic beverages, beverages
Level of use	Herb: non-alcoholic beverages: 360 ppm, alcoholic beverages 2400 ppm Extract: baked goods 171.3 ppm, frozen dairy 175 ppm, soft candy 171,3 ppm, gelatin pudding 175 ppm, non-alcoholic beverages 49.79 ppm, alcoholic beverages 46.06 ppm (9). Alcoholic beverages 15-10000 ppm, non-alcoholic: 1.5-2000 ppm, food: 10-120 ppm (IOFI 1995). Essential oil: baked goods 58.79 ppm, frozen dairy 59.64 ppm, soft candy 58.44 ppm, gelatin pudding 60.00 ppm, non-alcoholic beverages 18.12 ppm, alcoholic beverages 17.80 ppm, hard candy 0.77 ppm (9). Frozen dairy desserts 20 ppm, candy 20 ppm (IOFI 1996). Alcoholic beverages 18 ppm and non-alcoholic 1ppm (IOFI 1995).
Preparation	Alcoholic extract, essential oil, infusion, tincture
Main toxicological data	Thujone (α + β): Acute toxicity: toxic doses induced epileptic convulsions with general vasodolatation, decrease of blood pressure, lower cardiac rhythm and increase of respiratory amplitude. Short-term studies: 14 weeks gavage studies (isomeric composition of the material not specified): NOEL = 10 mg/kg/d for male rat and = 5 mg/kg/d for female rat. 13 weeks gavage studies with commercial mixture of α- and β-thujones: convulsive ED50 = 35.5 mg/kg/d for male rat and = 26.3 mg/kg/d for female rat; NOEL = 12.5 mg/kg/d for male rat and NOEL cannot be established for female rat. JECFA: ADI not specified (10). After oral administration of thujones (as a mixture of α- and β-isomers in the ratio of 9/2) to rabbits, 2 neutral urinary metabolites were obtained: *neo*-isothujanol (=β-hydroxy-α-thujane) and thujanol (=β-hydroxy-β-thujane) (11,12)
Data needed	28-day oral study with a definite isomeric composition of the mixture of thujone and mutagenicity studies. Information on utilisation of the sabinyl chemotype
Specific observations	None
Classification and limits	**Herb, essential oil and extracts: category 5 (with limits on camphor and thujone)**

National/int. evaluation	Thujone use restricted as in Annex II of Council Directive 88/388/EEC. Regulatory status in USA: Artemisia (Wormwood) FEMA 3114: CFR 172.510: finished food must be thujone free Artemisia (Wormwood Extract) FEMA 3115: CFR 172.510: finished food must be thujone free Artemisia (Wormwood oil) FEMA 3116: CFR 172.510: finished food must be thujone free
Main references	(1) Phytochemistry, *18*, 1591, (1979) (2) Dictionary of Natural Products, Volume 7, Chapman & Hall, (London), 417, (1993) (3) Fitoterapia, *60*, 460, (1989) (4) Z. Lebensmitt. Unters. Forsch., *176*, 363, (1983) (5) Z. Naturforsch., *36C*, 369, (1981) (6) Planta Med., *54*, 93, (1988) (7) Perfumer Flav., *17(2)*, 39, (1992) (8) J. Essent. Oil Res., *4*, 487, (1992) (9) Fenaroli (1995) (10) JECFA 25th meeting (1981) (11) Xenobiotica, *19*, 843, (1989) (12) Thujone, manuscript CoE (1999)
Data bases used	Chemical Abstracts Toxline Medline

Artemisia dracunculus L.

CE No.	64
Steinmetz No.	138
FEMA No.	Tarragon: 3943; Tarragon oil 2412
Order	Campanulales
Family	Asteraceae
Name	E Tarragon
	F Estragon, dragon
	D Estragon, Dragon
	I Estragone, targone, dragone
Synonyms	
Parts used	Herb
Important constituents	Herb: polyacetylenic compounds [capillene = agropyrene; 5-phenyl-1,3-pentadiyne; 4,6-heptadiyne-1,3-diol (acetylenic alcohol) and its glucoside; capillone and dehydrofalcarinone (polyacetylenic ketones)]; coumarinic and isocoumarinic derivatives [artemidinol; artemidiol; scopoletine; herciarine; 3-(1-butenyl)-1-H-2-benzopyran-1-one (E) and (Z)-form; 1-oxo-1-H-2-benzopyran-3-carboxaldehyde]; alcohols [9-hydroxygeraniol; 4-methoxybenzyl alcohol; 6-methyl-2-methylene-6-octene-1,3,8-triol]; aldehyde [3-(4-methoxyphenyl)-2-propenal] (1) Essential oil from French Tarragon or Italian Tarragon: terpenic hydrocarbons [α-pinene 0.55-1.99%; camphene 0.01-0.35%; β-pinene + sabinene 24.16-46.78%; myrcene 0.65-1.70%; limonene 0.60-4.65%; cis-ocimene 0.48-4.22%; trans-ocimene 0.35-6.59%]; phenylpropanoid derivatives [methyl chavicol (estragole) 60.46-81.90%; methyl eugenol 0.35-1.52; eugenol 0.05-0.60%]; + 1,8-cineole 6.88%; esters [cinnamyl and bornyl acetate 0.14%], [α-thujone 0.09%; β-thujone 0.09%]; p-methoxybenzaldehyde 0.11% (italian tarragon) (2,3,4,5,6) Essential oil from Russian Tarragon or German Tarragon: terpenic hydrocarbons [α-pinene 0.23-1.29%; camphene 0.04-0.22%; β-pinene 0.15-0.55%; sabinene 11.07-46.96%; myrcene 0.10-0.19%; limonene 1.71-3.52%; cis-ocimene 1.78-9.49%; trans-ocimene 4.68-9.10%]; phenylpropanoid derivatives [methyl chavicol (estragole) not detected-0.31%; methyl eugenol 5.86-35.59%; elemicin 0.49-59.69%; methyl isoeugenol 0.03-2.92%; trans-isoelimicin 0.26-17.73%] (2,3,4) Tarragon extract: polyacetylenic compounds [capillene; capillone; and dehydrofalcarinone derivatives] (7); alcohols [nerolidol; spathulenol] (7); terpenic hydrocarbons [myrcene 0.1%; limonene 0.8%; (Z)-β-ocimene 3.2%; (E)-β-ocimene + γ-terpinene 3.3%] (7);

	alkenyl benzene [methylchavicol 86.4%; methyleugenol 0.8%; eugenol 2.1%; elimicin] (8)
Active principles	Elimicin, estragole, methyleugenol, polyacetylene compounds, thujone
Other chemical components	Not known
Products in which used	Baked goods, condiments, meat products, candy, beverages, fats and oils, ices
Level of use	Herb: fats, oils 1535 ppm; meat products 682.4 ppm; condiments, relish 2731 ppm; Soft candy 0.28 ppm; soups 2232 ppm; non-alcoholic beverages 1.3 ppm; gravies 500 ppm (9). Essential oil: baked goods 414.2 ppm; fats, oils 41.4 ppm; frozen dairy 146.6 ppm; meat products 104.3 ppm; fish products 0.15 ppm; condiments, relish 95.33 ppm; soft candy 355.9 ppm; gelatin, pudding 156.6 ppm; non-alcoholic beverages 154.6 ppm; gravies 20 ppm (9). Alcoholic beverages: 260 ppm, food: 3-30000 ppm (IOFI 1995) Alcoholic extract: alcoholic beverages 50-250; non-alcoholic beverages up to 5000 (IOFI 1995)
Preparation	Decoction, infusion, tincture (20% in 65° alcohol), essential oil, distillate
Main toxicological data	Tarragon essential oil: No mutagenic activity in the Ames test, but induced DNA-damaging effect in *Bacillus subtilis* rec-assay (10) Estragole: hepatocarcinogenic for female mice when administered in the diet at levels of 0.23% (345 mg/kg bw) and 0.46% for 12 months (11,12). Ames test generally negative or weakly negative (13). At low doses, estragole mainly undergoes detoxication reactions, notably O-demethylation and side-chain cleavage, but as the dose is increased, the extent of O-demethylation falls and other pathways, notably 1'-hydroxylation come into prominence (14,15). This 1'-hydroxylation followed by esterification (by metabolic conjugation) could give rise to a carbonium ion as the ultimate carcinogenic metabolite. 1'-hydroxyestragole reacts with guanine and adenine residues of murine hepatic DNA *in vitro* or *in vivo,* to form specific adducts, as a consequence of its metabolic esterification with sulfate (16,17). Induced unscheduled DNA synthesis [UDS] in rat hepatocytes (18). Ames tests positive with acetyl ester of 1'-hydroxyestragole (11) and estragole 2',3' oxide (19) Methyleugenol: not mutagenic with Ames test (18,19) and with *Escherichia coli* WP2 reversion test (19). Positive in the yeast assay, in the *Bacillus subtilis* DNA repair test (18), and in UDS (20). In mice treated with methyleugenol, presence of DNA adducts in liver (16). Hepotocarcinogenic effects occurred in pre-weaning mice treated with methyleugenol or with its 1'-hydroxy metabolite by i.p; injection (12) Many metabolic pathways: allylic hydroxylation and oxidation of the side chain, O-demethylation, hydroxylation of the benzene ring (21). Ames test and UDS positive with 2,3-epoxy-methyleugenol

	(19). JECFA: no ADI specified (22). JECFA ADI for eugenol 0-2.5 mg/kg bodyweight and day (23)
Data needed	No data required
Specific observations	At least two varieties exist, *Artemisia dracunculus L* which is French tarragon and *Artemisia dracunculoides* Pursh. syn. *Artemisia dracunculus dracunculoides* syn. *Artemisia dracunculus var. inodora* which is Russian tarragon
Classification and limits	**Herb: category 3 (with limits on elimicin, estragole, methyleugenol, polyacetylene compounds, thujone)** **Essential oil and alcoholic extracts: category 4 (with limits on elimicin, estragole, methyleugenol, polyacetylene compounds, thujone)**
National/int. evaluation	Thujone use restricted as in Annex II of Council Directive 88/388/EEC. Methyleugenol: NTP (National Toxicology Program): long-term and carcinogenis studies on going Regulatory status in USA: tarragon: CFR 182.10 -582.10; FEMA No. 3043; Tarragon Extract: CFR 12.20; Tarragon oil: CFR 182.10 – 582.20; FEMA No. 2412
Main references	(1) Handbook of Natural Products, Chapman & Hall, Volume 7, 418, (1993) (2) Z. Lebensmitt. Unters. Forsch., 173, 365, (1981) (3) Perfumer Flav., *13(2)*, 50, (1988) (4) Perfumer Flav., *15(2)*, 75, (1990) (5) J. Essent. Oil Res., *1*, 111, (1989) (6) Planta Med., *57*, 237, and 450, (1991) (7) Phytochemistry, *18*, 1319, (1979) (8) Perf. Flav., *20(4)*, 38, (1995) (9) Fenaroli (1995) (10) Planta Med., *57*, 237, (1991) (11) J. Nat. Cancer Inst., *57*, 1323, (1976) (12) Cancer Res., *43*, 1124, (1983) (13) Mutat. Res., *60*, 143, (1979) (14) Biochem. Pharmacol., *30*, 1383, (1981) (15) Food Chem. Toxicol., *25*, 799, (1987) (16) Cancer Res., *36*, 1686, (1976); *41*, 176, (1981); and *45*, 3096, (1985) (17) Carcinogenesis, *5*, 1613, and 1623, (1984) (18) Environ. Mutagen., *8*, 1, (1986) (19) C.R. Soc. Biol., *171*, 1041, (1977) (20) Food Chem. Toxicol., *28*, 537, (1990) (21) CRC Handbook of Mammalian Metabolism of Plant Compounds, CRC Press, 68, (1993) (22) JECFA 25th meeting (1981) (23) JECFA 26th meeting (1982)
Data bases used	Chemical Abstracts Toxline

Artemisia spicata Wulf.

CE No.	66
Steinmetz No.	
FEMA No.	
Order	Campanulales
Family	Asteraceae
Name	E Geneppi, Black Juniper
	F Genépi mâle, Genépi noir
	D Schwarze Edelraute
	I Genepi maschio
Synonyms	Artemisia genepi Weber
Parts used	Herb
Important constituents	Herb (qualitative composition): terpenic hydrocarbons: mainly β-pinene; oxygenated terpenic compounds: [1,8 cineole (=eucalyptol); α- and β-thujones]; dihydrofuranoketone [davanone] (1); sesquiterpenic lactones [custinolide and artemorin (main bitter components); dehydroartemorin; santamarine; reynosin] (2)
	Essential oil: (max. levels in %): terpenic hydrocarbons: [mainly β-pinene 17.9%; α-pinene 1.2%; p-cymene 3.1%; γ-terpinene 2.8%; terpinolene; camphene; sabinene; limonene; myrcene; γ-elemene; β-farnesene]; oxygenated terpenic compounds: [mainly α-thujone 26%; β-thujone 6.8% terpinene-4-ol 12.2%; valeranone 3.3%; spathulenol 1.7%; 1,8-cineole; myrtenol; cis-sabinol]; esters [mainly sabinyl isobutyrate 1.0%; sabinyl isovalerate 2.6%; sabinyl n-valerate 1.5%; sabinyl acetate; α-terpenyl acetate; linalyl acetate; geranyl valerate] (1)
Active principles	Thujone
Other chemical components	Eucalyptol
Products in which used	Liqueurs (limited use because of the rarety and the high cost of the plant)
Level of use	Alcoholic extract of herb or flowers: alcoholic beverages: 30 ppm (IOFI,1995)
Preparation	Alcoholic extract
Main toxicological data	Thujone (α + β): Acute toxicity: toxic doses induced epileptic convulsions with general vasodilatation, decrease of blood pressure, lower cardiac rhythm and increase of respiratory amplitude (JECFA 1981). Short-term studies: 14 weeks gavage studies (isomeric composition of the material not specified): NOEL = 10 mg/kg/d for male

rat and = 5 mg/kg/d for female rat (JECFA, 1981). 13 weeks gavage studies with commercial mixture of α- and β-thujones: convulsive ED50 = 35.5 mg/kg/d for male rat and = 26.3 mg/kg/d for female rat; NOEL = 12.5 mg/kg/d for male rat and NOEL cannot be established for female rat (3)

After oral administration of thujones (as a mixture of α- and β-isomers in the ratio of 9/2) to rabbits, 2 neutral urinary metabolites were obtained: neoisothujanol (=β-hydroxy-α-thujane) and thujanol (=β-hydroxy-β-thujane) (4)

Eucalyptol (1,8-cineole): main urinary metabolites in brushtail possum (*Trichosurus vulpecula*) = 9-hydroxycineole and cineol-9-oic acid (5,6); in rats treated by gavage main metabolites = 2-hydroxycineole, 3-hydroxycineole (neutral metabolites) and 1,8-dihydroxy-10-carboxy-p-menthane (acidic metabolite) (7) [formation of this acidic metabolite requires the hydrolysis of the cyclic ether linkage]; in rabbits given eucalyptol by gavage same urinary neutral hydroxylated metabolites: 2- and 3-hydroxycineole (8)

Mutagenicity tests negative with *Bacillus subtilis* rec-assay (9), with CHO cells (no increase in SCE's) (10)

Costunolide: cytotoxic (11) and allergenic (12)

Data needed	No data required
Specific observations	None
Classification and limits	**Herb and alcoholic extract: category 4 (with limits on eucalyptol and thujone)**
National/int. evaluation	Thujone use restricted as in Annex II of Council Directive 88/388/EEC. FDA: not authorised in USA
Main references	(1) Z. Lebensm. Unters. Forsch., 175, 182, (1982) (2) J. Agric. Food Chem., 30, 518, (1982) (3) JECFA 25th meeting (1981) (4) Xenobiotica, 19, 843, (1989) (5) Xenobiotica, 10, 17, (1980) (6) Bull. Environ. Contam. Toxicol., 37, 759, (1986) (7) J. Agric. Food Chem., 37, 222, (1989) (8) J. Osaka City Med. Cent., 34, 267, (1986) (9) Mutat. Res., 226, 103, (1989) (10) J. Pharm. Sci., 58, 877, (1969) (11) Planta Med., 38, 289, (1980)
Data bases used	Chemical Abstracts Toxline

Artemisia umbelliformis Lam.

CE No.	68
Steinmetz No.	141
FEMA No.	
Order	Campanulales
Family	Asteraceae
Name	E Alpine wormwood
	F Armoise des Alpes
	D Edelraute, Alpenbeifuß, weißer Genipp.
	I
Synonyms	Artemisia mutellina Vill.
Parts used	Herb
Important constituents	Herb: umbellifolide
	Essential oil: (max. levels): terpenic hydrocarbons [α- and β-pinenes; sabinene; p-cymene; calamenene; δ-cadinene]; oxygenated terpenic compounds (main components) (~ 71.3%) [α-thujone 57.7%; β-thujone 8,6%; terpinen-4-ol 1.0%; spathulenol 2.2%]; esters [sabinyl isobutyrate 1.3%; sabinyl isovalerate 2.7%; sabinyl n-valerate 1.0%; bornyl acetate; sabinyl acetate; sabinyl propionate; linalyl caproate] (1)
Active principles	Thujone
Other chemical components	Not known
Products in which used	Alcoholic beverages
Level of use	Alcoholic extract: alcoholic beverages: 60 mg/l (IOFI 1995)
	Essential oil: not found
Preparation	Alcoholic extract
Main toxicological data	Thujones (α + β): Acute toxicity: toxic doses induced epileptic convulsions with general vasodolatation, decrease of blood pressure, lower cardiac rhythm and increase of respiratory amplitude. Short-term studies: 14 weeks gavage studies (isomeric composition of the material not specified): NOEL = 10 mg/kg/d for male rat and = 5 mg/kg/d for female rat. 13 weeks gavage studies with commercial mixture of α- and β-thujones: convulsive ED50 = 35.5 mg/kg/d for male rat and = 26.3 mg/kg/d for female rat; NOEL = 12.5 mg/kg/d for male rat and NOEL cannot be established for female rat. JECFA: ADI not specified (2)
	After oral administration of thujones (as a mixture of α- and β-isomers in the ratio of 9/2) to rabbits, 2 neutral urinary metabolites

	were obtained: neo-isothujanol (=β-hydroxy-α-thujane) and thujanol (=β-hydroxy-β-thujane) (3,4)
Data needed	No data required
Specific observations	Limited use because of the rarety and the high cost of the plant
Classification and limits	**Herb and alcoholic extracts: category 4 (with limits on thujone)**
National/int. evaluation	Thujone use restricted as in Annex II of Council Directive 88/388/EEC. FDA: not authorised in USA
Main references	(1) Z. Lebens. Unters. Forsch., *175*, 182, (1982) (2) JECFA 25th meeting (1981) (3) Xenobiotica, *19*, 843, (1989) (4) Thujone, manuscript CoE (1999)
Data bases used	Chemical Abstracts Toxline

Artemisia pallens

CE No.	69
Steinmetz No.	
FEMA No.	
Order	Campanulales
Family	Asteraceae
Name	E Davana
	F Davana
	D
	I
Synonyms	

Parts used	Herb
Important constituents	Herb: dihydrofurano compounds: [davanafuran; 2,5-dihydro-2-(5-hydroxy-1,5-dimethyl-2-oxo-3-hexenyl)-4-methyl-3-(3-methyl-2-butenyl)furan O-cinnamoyl; α- and β-dihydrorosefuran; 2,5-dihydro-2-hydroxy- 3-methyl-2-(3-methyl-2-butenyl)furan; davanone; 2-hydroxy-*iso*-davanone; 2-hydroperoxy-*iso*-davanone; cis-and trans-3-hydroxy-*allo*-davanone; 3,4-epoxy-2-hydroxy-*iso*-davanone; *cis*- and *trans*-3,4-epoxy-*allo*-davanone; artedouglasioxide] (1) Essential oil: terpenic hydrocarbons 8.55% [mainly bicyclo-germacrene 6.0%; bicycloelemene 0.3%; aromadendrene 0.23%; *allo*-aromadendrene 0.5%; p-cymene 0.8%; γ-cadinene 0.8%]; dihydrofurano compounds [*trans*-davanafuran 1.00%; and 2α-(3-methylbut-2-enyl)-4-methyl 2,5-dihydrofuran 1.50% (responsible for the characteristic odour of the oil); 2β-(3-methylbut-2-enyl)-4-methyl 2,5-dihydrofuran 1.00%; 2-(3-methylbut-2-enyl)-5-(5-cinnamoyl-oxy-2-oxo-1,5-dimethylhex-3-enyl)-3-methyl-2,5-dihy-drofuran; *cis*-davanone 38%; *trans*-davanone 5%; *iso*-davanone 3%; *nor*-davanone 1.5%; lavender lactone 3.0%; davanols 1.4% (two isomers); *cis*-hydroxydavanone 3.0%; *trans*-hydroxydavanone 2.0%; hydroxydavana ketone; *nor*-davana ether; davana ether 1.5%]; terpenic alcohols [geraniol 5.0%; nerol 10%; T-cadinol 2.9%; linalool 0.5%; borneol 0.5%; linalool oxide 0.5%; β-eudesmol 2.0%]; alkenylbenzene [eugenol 3.14%]; acids [mainly davanic acids (four isomers) 0.57-2.5%; *cis*- and *trans*-cinnamic acids 0.29%]; esters [ethyl davanate 0.7%; cinnamyl cinnamate 2.0%] (2,3,4,5)
Active principles	Not known
Other chemical components	Not known

Products in which used	Alcoholic and non-alcoholic beverages; food
Level of use	Essential oil: alcoholic beverages 2-8 mg/l; non-alcoholic beverages 1 mg/l; food 2-15 mg/kg (IOFI, 1995)
Preparation	Essential oil
Main toxicological data	Eugenol JECFA ADI 0-2.5 mg/kg body weight and day (6) Linalool: JECFA group ADI 0-0.5 mg/kg body weight and day (7)
Data needed	28 days toxicity and mutagenicity studies on essential oil
Specific observations	None
Classification and limits	**Herb and preparations: category 2**
National/int. evaluation	None
Main references	(1) Dictionary of Natural Products, Chapman & Hall, Volume 7, 419 (1993) (2) Perf. Flav., 5 *(April/May)*, 23, (1980) (3) Perf. Flav., *13(3)*, 50, (1988) (4) Phytochemistry, *30*, 549, (1991) (5) Perf. Flav., *20(1)*, 54, (1995) (6) JECFA 26th meeting (1982) (7) JECFA 23rd meeting (1979)
Data bases used	Chemical Abstracts Toxline

Artemisia pontica L.

CE No.	70
Steinmetz No.	142
FEMA No.	
Order	Campanulales
Family	Asteraceae
Name	E Pontique wormwood, Roman mugwort
	F Armoise pontique, Petite absinthe
	D Pontischer Absinth, Römischer Wermut
	I Assenzio gentile
Synonyms	
Parts used	Herb
Important constituents	Herb (direct head-space analysis) terpenic hydrocarbons [[α-pinene 0.2%; β-pinene > 0.1%; sabinene > 0.1%; p-cymene > 0.1%; myrcene > 0.1%; limonene > 0.1%; α-phellandrene > 0.1%]; sesquiterpene lactone [artemisia ketone 46%]; oxygenated terpenic compounds [α-thujone 14.3%; β-thujone 1.8%; 1,8-cineole 22.8%] (1) Essential oil: terpenic hydrocarbons [α- and β-pinenes ~ 0.1%; p-cymene 1.3%; limonene 2.9%; caryophyllene 1.0%; α-copaene 0.4%; sabinene ~0.1%; myrcene ~0.1%; α-phellandrene 1.0%; γ-terpinene 0.7%; γ-cadinene 1.4%; β-bourbonene 0.5%]; sesquiterpene lactone [artemisia ketone 22.8-46.2%; pontica-epoxide]; oxygenated terpenic compounds [α-thujone 13.5-30%; β-thujone 3.3-4.2%; 1,8-cineole (=eucalyptol) 12.0-23.0%; geraniol > 0.1%; eugenol 0.1%; Ho-trienol 1.8%]; phenolic derivatives [thymol 0.6%; carvacrol 0.2%] (1,2,3,4)
Active principles	Thujone
Other chemical components	Eucalyptol
Products in which used	Alcoholic and non-alcoholic beverages
Level of use	Alcoholic extract: alcoholic beverages 2-10000 mg/l; non-alcoholic beverages 2 ppm foods: < 7 ppm (IOFI, 1995) Essential oil: not found
Preparation	Alcoholic extract
Main toxicological data	Thujone (α + β): Acute toxicity: toxic doses induced epileptic convulsions with general vasodilatation, decrease of blood pressure, lower cardiac rhythm and increase of respiratory amplitude. Short-term studies: 14 weeks gavage studies (isomeric composition of the

material not specified): NOEL = 10 mg/kg/d for male rat and = 5 mg/kg/d for female rat. 13 weeks gavage studies with commercial mixture of α- and β-thujones: convulsive ED50 = 35.5 mg/kg/d for male rat and = 26.3 mg/kg/d for female rat; NOEL = 12.5 mg/kg/d for male rat and NOEL cannot be established for female rat. JECFA: ADI not specified (5)

After oral administration of thujones (as a mixture of α- and β-isomers in the ratio of 9/2) to rabbits, 2 neutral urinary metabolites were obtained: neo-isothujanol (=β-hydroxy-α-thujane) and thujanol (=β-hydroxy-β-thujane) (6,7)

Eucalyptol (1,8-cineole): main urinary metabolites in brushtail possum (*Trichosurus vulpecula*) = 9-hydroxycineole and cineol-9-oic acid (8,9); in rats treated by gavage main metabolites = 2-hydroxycineole, 3-hydroxycineole (neutral metabolites) and 1,8-dihydroxy-10-carboxy-p-menthane (acidic metabolite) (10) [formation of this acidic metabolite requires the hydrolysis of the cyclic ether linkage]; in rabbits given eucalyptol by gavage same urinary neutral hydroxylated metabolites: 2- and 3-hydroxycineole (11)

Mutagenicity tests negative with *Bacillus subtilis* rec-assay (12), with CHO cells (no increase in SCE's) (13)

Eugenol: JECFA ADI 0-2.5 mg/kg bodyweight and day (14)

Data needed	No data required
Specific observations	None
Classification and limits	**Herb and alcoholic extracts: category 4 (with limits on eucalyptol and thujone)**
National/int. evaluation	Thujone use restricted as in Annex II of Council Directive 88/388/EEC FDA: thujones: not authorised in USA
Main references	(1) High Resol. Chromatogr. Chromatogr. Commun., *5,* 182, (1982) (2) Rivista ital. EPPOS, *62,* 350, (1981) (3) Rivista ital. EPPOS, *7,* 409, (1966) (4) Planta Med., *31,* 97, (1977) (5) JECFA 25th meeting (1981) (6) Xenobiotica, *19,* 843, (1989) (7) Thujone, manuscript CoE (1999) (8) Xenobiotica, *10,* 17, (1980) (9) Bull. Environ. Contam. Toxicol., *37,* 759, (1986) (10) J. Agric. Food Chem., *37,* 222, (1989) (11) J. Osaka City Med. Cent., *34,* 267, (1986) (12) Mutat. Res., *226,* 103, (1989) (13) J. Pharm. Sci., *58,* 877, (1969) (14) JECFA 26th meeting (1982)
Data bases used	Chemical Abstracts Toxline

Artemisia vallesiaca Lam.

CE No.	71
Steinmetz No.	143
FEMA No.	
Order	Campanulales
Family	Asteraceae
Name	E Mountain wormwood
	F Armoise du Valais
	D Schwarzer Genipp
	I
Synonyms	Artemisia vallesana Ali.
Parts used	Herb
Important constituents	Herb (chloroform extract): sesquiterpene lactones [germacranolides (herbolides A and D)]; + oleuropaldehyde (terpenic aldehyde) (1) Essential oil: (main components) terpenic hydrocarbons [camphene 5.1%; γ-cadinene 0.5%; tricyclene β-pinene; p-cymene; copaene; calamenene], oxygenated terpenic compounds [1,8-cineole (=eucalyptol) 17%; camphor 33,3%; borneol 10.7%; spathulenol 0.8% + small amounts of carvone; verbenone; α-cadinol; α-terpineol; myrtenal] + small amounts of alkenylbenzene deivatives [eugenol; anethole; estragole]; esters [amyl valerate; bornyl acetate, propioanate, isobutyrate and valerate; carvyl valerate]; sesquiterpenic lactone [artemisia ketone] (2)
Active principles	Estragole
Other chemical components	Camphor, eucalyptol
Products in which used	Alcoholic beverages
Level of use	Not found
Preparation	Essential oil
Main toxicological data	Camphor: acute poisoning reported in humans after ingestion of 6-10 g camphor as stimulant. Classified as very toxic compound, with probable human lethal dose in the range of 50-500 mg/kg bw. (3,4). No effect on fetal growth, viability or morphological development at dosers causing minor maternal toxicity in female rats (5). No carcinomas developed after 4.5-6 months of twice-weekly painting of mice with a mixture of cyclic terpenes including camphor (6). Injected i.p. into Strain A mice 3 times/week for 8 weeks: no increase in primary lung tumours (6). Major metabolic pathways involve oxidation at the C-3 and especially the C-5 positions, or

reduction of the keto group [review in (7)]. In humans, hydroxylation at the 3-, 5-, 8- and 9- position. 5- and 8- or 9-hydoxycamphor are further oxidised to ketone and carboxylic acid, the latter is conjugated with glucuronic acid (4)

Eucalyptol (1,8-cineole): main urinary metabolites in brushtail possum (*Trichosurus vulpecula*) = 9-hydroxycineole and cineol-9-oic acid (8,9); in rats treated by gavage main metabolites = 2-hydroxycineole, 3-hydroxycineole (neutral metabolites) and 1,8-dihydroxy-10-carboxy-p-menthane (acidic metabolite) (10) [formation of this acidic metabolite requires the hydrolysis of the cyclic ether linkage]; in rabbits given eucalyptol by gavage same urinary neutral hydroxylated metabolites: 2- and 3-hydroxycineole (11)

Mutagenicity tests negative with *Bacillus subtilis* rec-assay (12), with CHO cells (no increase in SCE's) (13)

Borneol: mutagenic effects: ames test negative (14,15), negative with *Escherichia* coli WP2 uvrA, but positive with *Bacillus subtilis* rec-assay at 10 mg per dish (maximum dose tested) (16). Enzyme inducer in dogs (17), and in rats (18). In man, dogs, rabbits, borneol undergoes glucuronic conjugation, and the glucuronides are rapidly excreted in urine (19)

Eugenol: JECFA ADI 0-2.5 mg/kg bodyweight and day (20)

Anethole: JECFA ADI 0-2 mg/kg body weight and day (21)

Data needed	Level of use
Specific observations	None
Classification and limits	**Herb and preparations: category 4 (with limits on camphor, estragole and eucalyptol)**
National/int. evaluation	NTP: Carcinogenic studies ongoing Borneol: Council of Europe: Category B, Part I, Blue Book, 4th Edition. 28 day oral toxicity requested. Limits: Beverage: 2 mg/kg; Food B (condiments, seasonings): 5 mg/kg; Exception B (condiments, seasonings): 10 mg/kg
Main references	(1) Fitoterapia, *64*, 286, (1993) (2) Z. Lebensm. Unters. Forsch., *172*, 457, (1981) (3) Pediatrics, *57*, 42, (1976) (4) Arch. Toxicol., *51*, 101, (1982) (5) Chem. Abstr., *118*, 228029g and 228030a, (1993) (6) Food Cosmet. Toxicol., *16(Suppl. 1)*, 665, (1978) (7) CRC Handbook of Mammalian Metabolism of Plants Compounds, CRC Press Inc., *109*, (1991) (8) Xenobiotica, *10*, 17, (1980) (9) Bull. Environ. Contam. Toxicol., *37*, 759, (1986) (10) J. Agric. Food Chem., *37*, 222, (1989) (11) J. Osaka City Med. Cent., *34*, 267, (1986) (12) Mutat. Res., *226*, 103, (1989) (13) J. Pharm. Sci., *58*, 877, (1969) (14) Develop. Toxicol. Environ. Sci., *2*, 249, (1977)

(15) Arch. Environ. Contam. Toxicol., *28*, 248, (1995)
(16) J. Osaka City Med. Cent., *34*, 267, (1986)
(17) J. Biol. Chem., *136*, 229, (1940)
(18) Biochem. J., *113*, 12P, (1969)
(19) Biochem. J., *28*, 131, (1934); 30, 799, (1936)
(20) JECFA 26th meeting (1982)
(21) JECFA 51st meeting (1998)

Data bases used   Chemical Abstracts
Toxline

Artemisia vulgaris L.

CE No.	72
Steinmetz No.	144
FEMA No.	
Order	Campanulales
Family	Asteraceae
Name	E Common mugwort
	F Armoise commune
	D Gemeiner Beifuß
	I Assenzio selvatico
Synonyms	Artemisia dubia Wall. − Artemisia grata Wall. − Artemisia indica Willd.
	Artemisia lavandulaefolia D.C − Artemisia leptostachya D.C.
	Artemisia officinalis Gater. − Artemisia paniculata Roxb.
	Artemisia parvifolia Wight − Artemisia roxburghiana Bess.
Parts used	Herb
Important constituents	Herb: (qualitative composition) sesquiterpenic lactones [vulgarin; vulgarole; α-amyrin; epoxyartemisia ketone] alcohols [α-fernenol = 9(11)-fernen-3-ol (3-β *form*); 12-tricosanol] polyacetylenic derivatives [1,7,9-heptadecatriene-11,13,15-triyne (*all-E*)-*form*; 3,5-tridecatriene-7,9,11-triyne (3*E*,5*Z*)-*form*; 2-decene-4,6,8-triynoic acid, (*Z*)-*form*, methyl ester; 5-(5(methyl-2-thienyl)-2-penten-4-ynoic acid, (*Z*)-*form*, methyl ester]; acids [1-hydroxy-2,4(18),11(13)-eudesmatrien-12-oic acid, 1α-*form*; 3-oxo-1,4,11(13)-eudesmatrien-12-oic acid] (1)
	Essential oil (max. levels %): terpenic hydrocarbons [α-pinene 0.3%; β-pinene 2.2%; sabinene 15.9% caryophyllene 1.5%; camphene 0.45%; limonene 0.53%; p-cymene 3.3%]; oxygenated terpenic compounds [α-thujone 56.3%; β-thujone 7.5%; 1,8-cineole (=eucalyptol) 26.8%; camphor 20%; borneol 18.5%; α-terpineol 2.21%; geraniol 1.5%; eugenol 0.92%; linalool 1.14%; piperitone]; ester [bornyl acetate 18%] sesquiterpene lactone [vulgarole] (2,3,4,5)
Active principles	Polyacetylene compounds, thujone
Other chemical components	Camphor, eucalyptol
Products in which used	Alcoholic and non-alcoholic beverages, foods
Level of use	Alcoholic extract of herb: 2-10000 ppm in alcoholic beverages; 2-40 ppm in beverages; 10-120 ppm in foods (IOFI, 1995): of flowers: 2 ppm in alcoholic and non-alcoholic beverages (IOFI, 1995)

Tincture of herb: 2000 ppm in alcoholic beverages (IOFI, 1995)
Essentail oil of herb: 15 ppm in alcoholic beverages; 0.2-0.6 ppm in beverages (IOFI, 1995)

Preparation Alcoholic extract, tincture and essential oil of herb and alcoholic extract from flowers

Main toxicological data Thujone ($\alpha + \beta$): acute toxicity: toxic doses induced epileptic convulsions with general vasodilatation, decrease of blood pressure, lower cardiac rhythm and increase of respiratory amplitude. Short-term studies: 14 weeks gavage studies (isomeric composition of the material not specified): NOEL = 10 mg/kg/d for male rat and = 5 mg/kg/d for female rat. 13 weeks gavage studies with commercial mixture of α- and β-thujones: convulsive ED50 = 35.5 mg/kg/d for male rat and = 26.3 mg/kg/d for female rat; NOEL = 12.5 mg/kg/d for male rat and NOEL cannot be established for female rat. JECFA: ADI not specified (6)

After oral administration of thujones (as a mixture of a- and b-isomers in the ratio of 9/2) to rabbits, 2 neutral urinary metabolites were obtained: neo-isothujanol (=β-hydroxy-α-thujane) and thujanol (=β-hydroxy-β-thujane) (7,8)

Eucalyptol (1,8-cineole): main urinary metabolites in brushtail possum (Trichosurus vulpecula) = 9-hydroxycineole and cineol-9-oic acid (9,10); in rats treated by gavage main metabolites = 2-hydroxycineole, 3-hydroxycineole (neutral metabolites) and 1,8-dihydroxy-10-carboxy-p-menthane (acidic metabolite) (11) [formation of this acidic metabolite requires the hydrolysis of the cyclic ether linkage]; in rabbits given eucalyptol by gavage same urinary neutral hydroxylated metabolites: 2- and 3-hydroxycineole (12)

Mutagenicity tests negative with Bacillus subtilis rec-assay (13), with CHO cells (no increase in SCE's) (14)

Camphor: Acute poisoning reported in humans after ingestion of 6-10 g camphor as stimulant. Classified as very toxic compound, with probable human lethal dose in the range of 50-500 mg/kg bw. (15,16). No effect on fetal growth, viability or morphological development at doses causing minor maternal toxicity in female rats (17). No carcinomas developed after 4.5-6 months of twice-weekly painting of mice with a mixture of cyclic terpenes including camphor (18). Injected i.p. into Strain A mice 3 times/week for 8 weeks: no increase in primary lung tumours (18). Major metabolic pathways involve oxidation at the C-3 and especially the C-5 positions, or reduction of the keto group [review in (19)]. In humans, hydroxylation at the 3-, 5-, 8- and 9- position. 5- and 8- or 9-hydoxycamphor are further oxidised to ketone and carboxylic acid, the latter is conjugated with glucuronic acid (16)

Borneol: Mutagenic effects: Ames test negative (20,21), negative with Escherichia coli WP2 uvrA, but positive with Bacillus subtilus rec-assay at 10 mg per dish (maximum dose tested) (22). Enzyme inducer in dogs (23), and in rats (24). In man, dogs, rabbits, borneol undergoes glucuronic conjugation, and the glucuronides are rapidly

excreted in urine (25)
Eugenol: JECFA ADI 0-2.5 mg/kg bodyweight and day (26)
Linalool: JECFA group ADI 0-0.5 mg/kg bodyweight and day (27)

Data needed	No data required
Specific observations	None
Classification and limits	**Herb and preparations: category 4 (with limits on camphor, eucalyptol, polyacetylene compounds, and thujone)**
National/int. evaluation	Thujone use restricted as in Annex II of Council Directive 88/388/EEC. FDA: not authorised in USA Camphor: NTP: Carcinogenic studies ongoing Borneol: Council of Europe: Category B, Part I, Blue Book, 4th Edition. 28 day oral toxicity requested.
Main references	(1) Dictionary of Natural Products, Chapman & Hall, Volume 7, 420, (1993) (2) Planta Med., *30*, 211, (1976) (3) Z. Naturforsch., Teil C., *37*, 152, (1982) (4) Ann. Pharm. Fr., *43*, 397, (1985) (5) J. Nat. Prod., *49*, 941, (1986) (6) JECFA 25th meeting (1981) (7) Xenobiotica, *19*, 843, (1989) (8) Thujone, manuscript CoE (1999) (9) Xenobiotica, *10*, 17, (1980) (10) Bull. Environ. Contam. Toxicol., *37*, 759, (1986) (11) J. Agric. Food Chem., *37*, 222, (1989) (12) J. Osaka City Med. Cent., *34*, 267, (1986) (13) Mutat. Res., *226*, 103, (1989) (14) J. Pharm. Sci., *58*, 877, (1969) (15) Pediatrics, *57*, 42, (1976) (16) Arch. Toxicol., *51*, 101, (1982) (17) Chem. Abstr., *118*, 228029g and 228030a, (1993) (18) Food Cosmet. Toxicol., *16(Suppl. 1)*, 665, (1978) (19) CRC Handbook of Mammalian Metabolism of Plants Compounds, CRC Press Inc., 109, (1991) (20) Develop. Toxicol. Environ. Sci., *2*, 249, (1977) (21) Arch. Environ. Contam. Toxicol., *28*, 248, (1995) (22) J. Osaka City Med. Cent., *34*, 267, (1986) (23) J. Biol. Chem., *136*, 229, (1940) (24) Biochem. J., *113*, 12P, (1969) (25) Biochem. J., *28*, 131, (1934); *30*, 799, (1936) (26) JECFA 26th meeting (1982) (27) JECFA 23rd meeting (1979)
Data bases used	Chemical Abstracts Toxline

Artemisia herba-alba Asso.

CE No.	2011
Steinmetz No.	
FEMA No.	
Order	Campanulales
Family	Asteraceae
Name	E
	F
	D
	I
Synonyms	
Parts used	Herb
Important constituents	Herb: sesquiterpene lactones [eudesmanolides; germacranolides; herbolides; artemisia alcohol; artemisia adducts]; 2,6-dimethoxyphenol (1)

Essential oils from Morocco [many chemotypes]:

β-thujone chemotype: terpenic hydrocarbons [santolina-triene < 0.5%; tricyclene < 0.2%; a-pinene < 0.1%; camphene 0.4-7.5%; sabinene 2.1-2.4%; β-pinene < 0.5%; p-cymene 0.6-1.0%; γ-terpinene < 0.2%]; terpenic ketones [α-thujone 0.5-17.0%; β-thujone 43.4-94.0%; camphor 2.5-15.0%; pinocarvone < 1.3%]; alcohols [, 1,8-cineole (=eucalyptol) 1.8-5.8%; trans-pinocarveol < 1.0%; 1-terpinen-4-ol 0.3-2.4%; borneol < 1.1%]; esters [cis-chrysanthenyl acetate < 1.9%; (Z)-lyratyl acetate < 0.3%; bornyl acetate < 0.4%; sabinyl acetate < 0.5%]; chrysanthenone < 4.8% (2)

α-thujone chemotype: terpenic hydrocarbons [santolina-triene < 0.1%; tricyclene < 0.2%; α-pinene < 0.4%; camphene 1.4-3.7%; sabinene < 1.4%; β-pinene < 0.1%; p-cymene 0.3-1.4% γ-terpinene < 0.2%]; terpenic carbonyl derivatives [α-thujone 36.8-82.0%; β-thujone 6.0-16.2%; camphor 11.0-19.0%; pinocarvone < 0.5%; myrtenal < 0.3%]; alcohols [yomogi alcohol < 1.1%; artemisia alcohol < 0.6%; borneol 0.6-0.9%]; esters [santolinyl acetate < 0.4%; cis-chrysanthenyl acetate < 1.7%; (E)-lyratyl acetate < 0.5%; bornyl acetate < 0.3%; sabinyl acetate < 0.6%]; chrysanthenone < 3.3% (2)

Camphor chemotype: terpenic hydrocarbons [santolina-triene 0.1-1.2%; tricyclene < 0.5%; α-pinene < 0.5%; camphene 2.5-15.0%; sabinene 0.1-0.4%; β-pinene 0.1-0.4%; p-cymene 0.8-2.3%; γ-terpinene < 0.7%]; terpenic carbonyl compounds [α-thujone 2.5-25.0%; β-thujone 0.5-7.5%; camphor 40.0-70.0%; pinocarvone 1.6-3.5%; verbenone 0.4-0.8%; myrtenal 0.8-1.3%]; alcohols [1,8-

cineole (=eucalyptol) 2.6-15.0%; 1-terpinen-4-ol 0.6-2.1%; yomogi alcohol 1.2-2.5%; artemisia alcohol 1.3-4.4%; borneol 1.0-2.5%]; esters [santolinyl acetate 0.8-1.4%; cis-chrysanthenyl acetate < 0.2%; (E)-lyratyl acetate 0.4-0.5% (Z)-lyratyl acetate 0.5-0.8%; bornyl acetate 0.3-0.7%; sabinyl acetate < 0.6%]; chrysanthenone 0.1- 3.4% (2)

Chrysanthenone chemotype: (analysis of one sample): terpenic hydrocarbons [α-pinene 1.2%; camphene 2.0%; sabinene 0.7%; α-phellandrene 0.3%; p-cymene 0.9%; γ-terpinene 0.1%;a-copaene 0.3%]; ketones [α-thujone 2.9%; β-thujone 6.0%; camphor 7.2%; verbenone 0.1%; chrysanthenone 51.4%]; alcohols [trans-pinocarveol 0.2%; 1,8-cineole (=eucalyptol) 3.0%; 1-terpinen-4-ol 0.3%] esters [cis-chrysanthenyl acetate 5.5%; bornyl acetate 0.1%; sabinyl acetate < 0.6%] (2)

Davanone chemotype: terpenic hydrocarbons [α-pinene < 0.2%; camphene <4.2%; β-pinene < 0.4%; myrcene 0.2-1.4%; p-cymene 0.5-3.3%; β-phellandrene < 1.2%; α-copaene < 1.4%; γ-muurolene 0.3-1.3%; allo-aromadendrene < 0.3%]; carbonyl compounds [α-thujone 0.4-5.8%; β-thujone 0.2-5.0%; camphor trace-11.0%; chrysanthenone trace-10.0%; cis-jasmone 0.7-1.8%]; alcohols [1,8-cineole(=eucalyptol) < 2.0%; linalool 2.1-2.9%]; esters [cis-chrysanthenyl acetate < 2.9%; bornyl acetate < 0.2%]; dihydrofurano compounds [davanone 20.0-70.0%; nordavanone 0.7-1.5%; davana furane < 1.0%; davana ether < 3.0%; cis-methyl-5-vinyl-5-tetrahydrofuryl-2-methyl-ketone < 0.6%; trans-methyl-5-vinyl-5-tetrahydrofuryl-2-methyl-ketone < 0.6%] (2,3)

cis-chrysanthenyle acetate chemotype: hydrocarbons [camphene 1.5-2.5%; p-cymene; α-pinene; 1-methyl, 4-isopropenyl-benzene]; alcohols [carveol; cuminyl alcohol]; 1,8-cineole (=eucalyptol) 3.0-12.0%; carbonyl compounds [camphor 10.0-30.0%; chrysanthenone; cuminaldehyde; filifolone; piperitone; α-thujone traces; β-thujone traces-29%; cis-chrysanthenyle acetate 38.0-71.0% (3)

Essential oils from Israel [many chemotypes]:

1,8-cineole + α-thujone chemotype: terpenic hydrocarbons [p-cymene 2.1%; γ-terpinene 0.5%; α-thujene 0.4%; sabinene 0.4%; β-pinene 0.3%; β-cubebene 0.3%]; terpenic alcohols [1,8-cineole (=eucalyptol) 50.0%; terpinen-4-ol 3.5%; α-terpineol 0.5%; borneol 2.4%; trans-pinocarveol 0.7%; cis-sabinene hydrate 1.6%; myrtenol 0.3%]; terpenic ketones [α-thujone 27.0%; β-thujone 0.5%; camphor 3.0%]; aldehyde [myrtenal 0.3%] (4)

1,8-cineole + β-thujone chemotype: terpenic hydrocarbons [p-cymene 1.5%; sabinene 0.4%; camphene 0.5%; artemisia triene 0.8%]; terpenic alcohols [1,8-cineole (=eucalyptol) 13.0%; terpinen-4-ol 1.5%; α-terpineol 0.4%; piperitol 0.7%; borneol 11.0%; artemisia alcohol 10.0%; santolina alcohol 6.0% yomogi alcohol 8.8%; lyratol 5.7%]; terpenic ketones [α-thujone 4.2%; β-thujone 12.4%; camphor 9.0%; pinocarvone 0.3%; pinocarvone 0.3%]; esters [bornyl acetate 0.3%; terpenyl acetate 0.6%; cis-chrysanthenyl acetate 0.2%; myrtenyl acetate 1.5%] (4)

1,8-cineole + camphor chemotype: terpenic hydrocarbons [p-cymene 2%; α-thujene 0.1%; sabinene 0.3%; camphene 3.4%; β-pinene 0.4%; β-cubebene 2.2%; γ-elemene 0.3%]; terpenic alcohols [1,8-cineole (=euclayptole) 38.0%; terpinen-4-ol 5.0%; a-terpineol 0.9%; borneol 3.0%; *trans*-pinocarveol 0.8%; *cis*-sabinene hydrate 1.9%; artemisia alcohol 0.4%; santolina alcohol 0.7%; yomogi alcohol 0.3%; lyratol 3.4%]; terpenic ketones [α-thujone 1.4%; β-thujone 0.7%; camphor 25.0%; pinocarvone 0.4%; camphenilone 1.0%]; aldehyde [myrtenal 0.4%]; esters [bornyl acetate 0.4%; *cis*-chrysanthenyl acetate 0.1%; myrtenyl acetate 0.4%] (4)

Chrysanthenol chemotype: terpenic hydrocarbons [p-cymene 0.3%; g-terpinene 0.5%; α-copaene 0.6%; g-elemene 1.2%; β-cubebene 2.8%]; terpenic alcohols [1,8-cineole 4.8%; terpinen-4-ol 0.5%; trans-pinocarveol 1.9%; *cis*-chrysanthenol 24.5%]; carbonyle compounds [pinocarvone 0.6%; xanthoxylin (=2-hydroxy-4,6-dimethoxy-acetophenone) 9.1%; *cis*-jasmone 1.1%; chrysanthenone 4.4%; camphor 0.1%; davanone 2.0%]; esters [*cis*-chrysanthenyl acetate 6.4%] (4)

cis-chrysanthenyl acetate chemotype: terpenic hydrocarbons [p-cymene 1.2%]; terpenic alcohols [1,8-cineole (=eucalyptol) 7.4%; terpinen-4-ol 3.5%; α-terpineol 0.5%; p-cymen-8-ol 0.6%; piperitol 0.2%; *cis*-chrysanthenol 7.1%; *trans*-pinocarveol 1.8%; myrtenol 0.3%]; terpenic ketones [camphor 0.7%; chrysanthenone 4.4%; davanone 0.6%; *cis*-jasmone 4.0%]; esters [bornyl acetate 0.3%; *cis*-chrysanthenyl acetate 25.0%] (4)

Essential from Spain: [1,8-cineole + camphor chemotype]: terpenic hydrocarbons [α-terpinene 0.3%; p-cymene 3.9%; camphene 1.9%; α-pinene 1.7%; β-pinene 0.8%; caryophyllene 0.7%; β-elemene 2.7%; α-guiaiene or β-cubebene 6.0%; δ-cadinene 0.7%]; terpenic alcohols [1,8-cineole (=euclayptole) 13.3%; terpinen-4-ol 4.8%; α-terpineol 6.3%; piperitol 0.5%; borneol 4.8%; p-cymen-8-ol 0.5%]; cabonyle compounds [camphor 15.0%; chrysanthenone 4.5%; cumin aldehyde 0.6%]; esters [bornyl acetate 0.3%; terpenyl acetate 0.4%; bornyl propionate 2.6%] (4)

Active principles	Thujone
Other chemical components	Camphor, eucalyptol
Products in which used	Alcoholic beverages
Level of use	Alcoholic extract: 200 ppm in alcoholic beverages
Preparation	Alcoholic extract Chemotypes used: not found
Main toxicological data	*Thujone* ($\alpha + \beta$): Acute toxicity: toxic doses induced epileptic convulsions with general vasodilatation, decrease of blood pressure, lower cardiac rhythm and increase of respiratory amplitude. Short-term studies: 14 weeks gavage studies (isomeric composition of the material not specified): NOEL = 10 mg/kg/d for male rat and = 5

mg/kg/d for female rat. 13 weeks gavage studies with commercial mixture of a- and b-thujones: convulsive ED50 = 35.5 mg/kg/d for male rat and = 26.3 mg/kg/d for female rat; NOEL = 12.5 mg/kg/d for male rat and NOEL cannot be established for female rat. JECFA: ADI not specified (6).

After oral administration of thujones (as a mixture of a- and b-isomers in the ratio of 9/2) to rabbits, 2 neutral urinary metabolites were obtained: neo-isothujanol (=β-hydroxy-α-thujane) and thujanol (=β-hydroxy-β-thujane) (7,8)

Eucalyptol (1,8-cineole): main urinary metabolites in brushtail possum (*Trichosurus vulpecula*) = 9-hydroxycineole and cineol-9-oic acid (9,10); in rats treated by gavage main metabolites = 2-hydroxycineole, 3-hydroxycineole (neutral metabolites) and 1,8-dihydroxy-10-carboxy-p-menthane (acidic metabolite) (11) [formation of this acidic metabolite requires the hydrolysis of the cyclic ether linkage]; in rabbits given eucalyptol by gavage same urinary neutral hydroxylated metabolites: 2- and 3-hydroxycineole (12)

Mutagenicity tests negative with *Bacillus subtilis* rec-assay (13), with CHO cells (no increase in SCE's) (14)

Camphor: acute poisoning reported in humans after ingestion of 6-10 g camphor as stimulant. Classified as very toxic compound, with probable human lethal dose in the range of 50-500 mg/kg bw. (15,16). No effect on fetal growth, viability or morphological development at doses causing minor maternal toxicity in female rats (17). No carcinomas developed after 4.5-6 months of twice-weekly painting of mice with a mixture of cyclic terpenes including camphor (18). Injected i.p. into Strain A mice 3 times/week for 8 weeks: no increase in primary lung tumours (18). Major metabolic pathways involve oxidation at the C-3 and especially the C-5 positions, or reduction of the keto group [review in (19)]. In humans, hydroxylation at the 3-, 5-, 8- and 9- position. 5- and 8- or 9-hydoxycamphor are further oxidised to ketone and carboxylic acid, the latter is conjugated with glucuronic acid (16)

Borneol: mutagenic effects: Ames test negative (20,21), negative with *Escherichia* coli WP2 uvrA, but positive with *Bacillus subtilis* rec-assay at 10 mg per dish (maximum dose tested) (22). Enzyme inducer in dogs (23), and in rats (24). In man, dogs, rabbits, borneol undergoes glucuronic conjugation, and the glucuronides are rapidly excreted in urine (25)

Linalool: JECFA group ADI 0-0.5 mg/kg bodyweight and day (27)

Data needed	No data required
Specific observations	None
Classification and limits	Herb and alcoholic extracts: category 4 (with limits on camphor, eucalyptol and thujone)
National/int. evaluation	Thujone use restricted as in Annex II of Council Directive 88/388/EEC.

Main references

(1) Dictionary of Natural Products, Chapman & Hall, Volume 7, 416, (1993)
(2) Artemisie – Ricerce ed Applicazione, 2 (Supplemento alla rivista " Quaderno Agricolo "), 131, (1985)
(3) Sciences Aliments, *2*, 515, (1982)
(4) Phytochemistry, *25*, 2343, (1986)
(5) Phytochemistry, *27*, 433, (1988)
(6) JECFA 25th meeting (1981)
(7) Xenobiotica, *19*, 843, (1989)
(8) Thujone, manuscript CoE (1999)
(9) Xenobiotica, *10*, 17, (1980)
(10) Bull. Environ. Contam. Toxicol., *37*, 759, (1986)
(11) J. Agric. Food Chem., *37*, 222, (1989)
(12) J. Osaka City Med. Cent., *34*, 267, (1986)
(13) Mutat. Res., *226*, 103, (1989)
(14) J. Pharm. Sci., *58*, 877, (1969)
(15) Pediatrics, *57*, 42, (1976)
(16) Arch. Toxicol., *51*, 101, (1982)
(17) Chem. Abstr., *118*, 228029g and 228030a, (1993)
(18) Food Cosmet. Toxicol., *16(Suppl. 1)*, 665, (1978)
(19) CRC Handbook of Mammalian Metabolism of Plants Compounds, CRC Press Inc., 109, (1991)
(20) Develop. Toxicol. Environ. Sci., *2*, 249, (1977)
(21) Arch. Environ. Contam. Toxicol., *28*, 248, (1995)
(22) J. Osaka City Med. Cent., *34*, 267, (1986)
(23) J. Biol. Chem., *136*, 229, (1940)
(24) Biochem. J., *113*, 12P, (1969)
(25) Biochem. J., *28*, 131, (1934); *30*, 799, (1936)
(26) JECFA 23rd meeting (1979)

Data bases used

Chemical Abstracts
Toxline

Aspidosperma quebracho-blanco Schlechtend.

CE No.	78
Steinmetz No.	163
FEMA No.	Bark extract: 2972
Order	Gentianales
Family	Apocynaceae
Name	E Quebracho blanco, White Quebracho, Aspidosperma
	F Quebracho
	D Quebrachobaum, weißer Quebracho
	I Quebracho
	SP Quebracho corteza, extracto
Synonyms	Note 'Red Quebracho' is obtained from a different species (1) Schinopsis quebracho-colorado (Schlecht.) Barkl. et T. Meyer
Parts used	Bark (1), (2), (3), wood (3)
Important constituents	Alkaloids (0.3 to 1.5%) of bark (1, 2): aspidospermine (30% of alkaloids), quebrachine (yohimbine) (10% of alkaloids), deacetylaspidospermine (5% of alkaloids), aspidospermatin (3% of alkaloids), aspidospermatidine (3% of alkaloids), 1-methylaspidospermatidine (0.5% of alkaloids), quebrachimine and quebrachit. Leaves contain rhazinilam (lactam) (2)
Active principles	Not known
Other chemical components	Not known
Products in which used	Bark extract used in flavouring non-alcoholic and alcoholic beverages, frozen dairy products, soft candy, baked goods (3), frozen dairy deserts, gelatin, puddings (1). Used in fruit, rum and vanilla flavourings (4)
Level of use	Bark extract, highest average maximum use level is about 0.003% reported in candy 29.8 ppm and baked goods 34.5 ppm (1). Also frozen dairy products 25 ppm; gelatin and puddings 24.89 ppm; non-alcoholic beverages 14.87 ppm and alcoholic beverages 14.97 ppm (3)
Preparation	Fluid extract, tincture, elixir (alcoholic extract) of bark (2), crude white quebracho (1)
Main toxicological data	No toxicological data on bark extract or individual alkaloids. Pharmacological data – Aspidospermine reported to antagonise the effect of acetylcholine on frog muscle in a dose dependent manner (0.5-2.2 mg), slow frog heart rate in situ (5 mg/ml solution), depress peristaltic movements of frog duodenum (0.1-0.25 mg) and reduce

	carotid arterial blood pressure in chloralosed cats (5-30 mg i.v.) (6). Rhazinilam has analgesic effect in the acetic acid writhing test in mice (2). Yohimbine is an 2 adrenoreceptor blocker and is used in the treatment of organic impotence but autonomic side effects produced at doses of 15-20 mg/day (7)
Data needed	Quantitative data on chemical components in bark and bark extracts and, if necessary, 28-day oral study and study on mutagenicity. Further toxicological and compositional data is required on the bark extract used in flavouring preparations
Specific observations	None
Classification and limits	**Bark: category 5** **Wood: category 5**
National/int. evaluation	UK FACC (1976) Bark in Appendix 2 Bark Extract is FEMA 2972
Main references	(1) Leung (1996) (2) J. of Pharm. Science USA, 62 11. (1973) (3) Fenaroli (1995) (4) MAFF (1995) (5) J. of Pharm. Sci USA, 62, 11 1989 (1973) (6) J. of pharm. and Pharmacol. 7 46-49 (1955) (7) The Lancet August 22, 421-43. (1987)
Data bases used	Chemical Abstracts (1967-91) FSTA (1969-91) Toxline, Toxlit, Toxnet (1981-91) Embase (1974-91) Biosis (1973-91) *Keywords:* Aspiosperma(w)quebracho, White(w) Quebracho, Quebracho (w) blanco, Quebracho bark extract

Berberis vulgaris L.

CE No.	86
Steinmetz No.	189
FEMA No.	
Order	Ranunculales
Family:	Berberidaceae
Name	E Barberry
	F Epine-vinette
	D Sauerdorn, Berberitze
	I Berbero
	SP Agracejo
Synonyms	Berberis thunbergii DC
	Mahonia aquifolium Nutt.
Parts used	Fruits
Important constituents	The fruit contains normal plant constituents. They do not contain alkaloids such as berberine which is found in the leaves and bark of the plant (1,3)
Active principles	Not known
Other chemical components	Not known
Products in which used	The fruits are used to prepare juices and jellies (2). B. vulgaris is not used for flavouring purposes according to the British survey
Level of use	Fruit: beverages; food 40 g/kg (IOFI, 1994)
Preparation	No data available
Main toxicological data	No relevant data found
Data needed	No data required
Specific observations	Fruit used as a foodstuff
Classification and limits	**Fruit: category 1**
National/int. evaluation	None
Main references	(1) List & Hörhammer (1967-80)
	(2) Duke (1986)
	(3) Poisonous plants in Britain and their effects on animal and man. Cooper & Johnson, MAFF ref. book 161, HMSO London (1984)
Data bases used	Chem Abs 1966-97
	Medline 1967-97
	Biosis 1967-97
	Keywords: Latin and English names

Boronia megastigma Nees ex Bartl.

CE No.	91
Steinmetz No.	-
FEMA No.	Boronia flowers absolute: 2167
Order	Rutales
Family	Rutaceae
Name	E Boronia
	F Boronia
	D Boronia, Korallenraute
	I Boronia
	SP Boronia, absoluto
Synonyms	-

Parts used	Flowers
Important constituents	Boronia concrete: terpenic ketone: β-ionone 11.9-22.5%; esters [dodecyl acetate 5.6-11.1%; methyl jasmonate 3.4-7.2%]; hydrocarbons [(Z) heptadec-8-ene 19.7-31.2%; α-pinene 2.3-11.0%; β-pinene 1.5-15.0%; limonene 1.0-2.7%]; linalool 0.9-1.7%; "sesquicineole" traces-19.3% (1)
	Boronia absolute (qualitative data): hydrocarbons [α-pinene; β-pinene; camphene; myrcene; limonene; ocimene; methyl naphtalene; pentadecane; heptadecene]; alcohols [linalool; dodecanol]; ketones [β-ionone; α-ionone; dihydro-β-ionone; menthone; 5,6-epoxyionone]; esters [dodecyl acetate; tetradecyl acetate; methyl decanoate; ethyl decanoate] (2)
Active principles	Not known
Other chemical components	Not known
Products in which used	Baked goods, frozen dairy, condiment, relish, candy, gelatin, puddings, alcoholic and non-alcoholic beverages
Level of use	Flowers: beverages 5 g/l, foods 10 g/kg, (IOFI 1994)
	Boronia absolute: baked goods 11.62 ppm; frozen dairy 7.12 ppm; condiment: relish 0.12 ppm; soft candy 9.39 ppm; gelatin, puddings 8.60 ppm; non-alcoholic beverages 1.76 ppm; alcoholic beverages 7.18 ppm (3)
Preparation	Boronia concrete and absolute
Main toxicological data	1-β-ionone: NOEL 10 mg/kg/day in rats (90 days study in diet). Group ADI for α- and β-ionone = 0-0.1 mg/kg7d, single or in combination (4)
	Three metabolic pathways: reduction of the keto goup, double

bond reduction and oxidation of the alicyclic ring in rabbits (5,6) 2-α-pinene: not mutagenic in Ames tests; weak tumour-promoting effect on mouse skin when applied after a single dose of known carcinogen; foetotoxic only at a maternally toxic dose level when administered orally to pregnant rats with several other terpenes; repeated administration at lower levels in rats caused enzyme induction. In man, has been given orally in combination with several other terpenes to treat gallstones (7)

Data needed	No data required
Specific observations	Flowers are used as a foodstuff
Classification and limits	**Flowers: category 2** **Concrete and absolute: category 2**
National/int. evaluation	Boronia flowers: CFR 172.510; Boronia flowers absolute: FEMA 2167
Main references	(1) Perfumer&Flavorist 8, p.3 (1983) (2) Perfumer&Flavorist 8, p.61 (1983) (3) Fenaroli 1, p.53 (1995) (4) JECFA, 28th meeting (1984) (5) Helv. Chim. Acta 33, p.1276 (1950) (6) Biochem. J. 119, p.281 (1970) (7) BIBRA Toxicity Profile: a-pinene (1992)
Data bases used	Chemical Abstracts (1965-96)

Boswellia sacra Flueckiger

CE No.	93
Steinmetz No.	201
FEMA No.	-
Order	Rutales
Family	Burseraceae
Name	E Olibanum frankincense, Olibanum tree
	F Oliban, encens frankincense, arbre à incens
	D Weihrauchbaum, Gummiharzbaum
	I Olibano, incenso, olibanum, arbrere di incesio
	SP Olibano
Synonyms	Boswellia carterii Birdw.
Parts used	Gommo-oleoresin
Important constituents	Gum-resin: gum 27-35%; resin 60-70%; essential oil 4-7%(1) Essential oil (yield 4-7%): terpenic hydrocarbons [α-pinene 5-43%; α-thuyene 61%; camphene 1-2%; β-pinene 1.5%; limonene 1.5-6%; sabinene 1%; p-cymene 7.5%]; alcohols and esters [octanol 8%; octyl acetate 1.5-52%; linalool 2.5%]; 1,8-cineol 2.7%; carbonyl compounds [verbenone 6.5%; campholenic aldehyde 1.5%]; α- and β-boswellic acid; traces of α-campolytic acid; α- + γ-campholenic acid; cis + trans10-thujanic acid (1,2,3)
Active principles	Not known
Other chemical components	Eucalyptol
Products in which used	Alcoholic and non-alcoholic beverages, dairy products, candy, baked goods, gelatin, puddings, meat products
Level of use	Baked goods 5.81 ppm; frozen dairy 2.39 ppm; meat products 11.22 ppm; non-alcoholic beverages 1.54 ppm; alcoholic beverages 6.57 ppm; gelatin, puddings 1.75 ppm; soft candy 3.66 ppm (4)
Preparation	Essential oil, resinoid
Main toxicological data	α-Pinene: BIBRA Toxicity Profile (5)
Data needed	28-day oral toxicity study of a-pinene and a-thuyene
Specific observations	None
Classification and limits	Gommo-oleoresin, essential oil and resinoid: category 5
National/int. evaluation	None

Main references   (1) Perfumer&Flavorist 7, p.48 (1982)
(2) Leung p.245 (1980)
(3) Perfumer&Flavorist 9, p.19 (1985)
(4) Fenaroli 1, p.201 (1995)
(5) BIBRA toxicity profile (1992)

Data bases used   Chemical Abstracts, Pascal (1965-95)
Keywords: Boswellia carterii, Olibanum

Bursera ssp.

CE No.	236
Steinmetz No.	579
FEMA No.	Linaloe wood oil: 2634
Order	Rutales
Family	Burseraceae
Name	**E** Linaloe tree
	F Linaloé (du Mexique), bursera
	D Mexikanischer Linaloenbaum, Bursera, Elemi
	I Linaloe
	SP Linaloe
Synonyms	-
Parts used	Wood
Important constituents	Essential oil of wood: esters [linalyl acetate 40-70%; geranyl actate 3.5%; neryl acetate 2.5%;]; alcohols [linalool 30-48%;α-terpineol 8.5%; geraniol 1%; methyl-heptanol 1.5%]; *cis*-linalool oxide-furanoid 2%: *trans*-linalool oxide-furanoid 1%; hydrocarbons [limonene 0.3%; myrcene 0.3%; and a mixture of *cis*- and *trans*-2,6,6-trimethyl-2-vinyl-5-acetoxytetrahydropyrans] + 8% of sesquiterpene waxes and resinous substances (1)
Active principles	Not known
Other chemical components	Not known
Products in which used	Candy, alcoholic and non-alcoholic beverages, dairy products, ice-cream and ices, baked goods, gelatin and puddings
Level of use	Linaloe wood oil (Bursera delpechiana Poiss. and other Bursera spp.): baked goods 27.45 ppm; frozen dairy 16.13 ppm; soft candy 20.75 ppm; gelatin, puddings 16.27 ppm; non-alcoholic beverages 8.66 ppm; alcoholic beverages 9.47 ppm (2)
Preparation	Essential oil
Main toxicological data	**Essential oil: not irritating, not sesitisation, not phototoxic (3)**
	2-Linalool and linalyl acetate: ADI 0-0.5mg/kg bw (4)
Data needed	No data required
Specific observations	None
Classification and limits	Wood and essential oil of wood: category 2
National/int. evaluation	Linaloe wood oil: CFR 172.510; FEMA No. 2634

Main references   (1) Perfumer&Flavorist 1:4 (1976)
                           (2) Fenaroli 1:177 (1995)
                           (3) Food Cosmet. Toxicol. 17:849 (1979)
                           (4) JECFA 23rd Session (1979)

Data bases used   Chemical Abstracts (1965-96)
                           Bursera fagaroides (H.B.K.) Engl.
                           Bursera glabrifolia (H.B.K.)
                           Bursera penicillata (Sessé&Moc.ex DL)spp.
                           Bursera delpechiana Poiss.
                           Bursera aleoxylon (Schiede) Engl.

Cananga odorata Hook. fil. et Thomson f. macrophylla

CE No.	103 b
Steinmetz No.	
FEMA N	Canaga oil: 2232
Order	Magnoliales
Family	Annonaceae
Name	E Cananga
	F Cananga des Molluques
	D Cananga
	I Cananga
	SP Cananga, aceite esencial
Synonyms	Canangium odoratum Baill. f. macrophylla
Parts used	Herb (leaf)
Important constituents	Leaf: terpenes [α-pinene 14%, β-pinene 4.5%, myrcene 0.6%, limonene 0.4%, δ-elemene 1.1%, α-ylangene 0.3%, α-copaene 2.9%, β-cubebene 0.7%, β-elemene 1.4%, β-caryophyllene 26.3%, α-amorphene 0.4%, α-humulene 6.3%, germacrene D 11.7%, α-muurolene 1.1%, bicyclogermacrene 0.6%, γ-cadinene 0.8%, δ-cadinene 3.0%]; oxygenated terpenic compounds [linalool 1.9%, α-terpineol 0.3%, caryophyllene epoxide 0.6%, T-cadinol 0.9%, α-cadinol 0.8%], aliphatic alcohols [n-hexanol 10.2%, (Z)-hex-3-enol 3.1%, (E)-hex-2-enol 0.2%]; (E)-hex-2-enal 2.0% (1,2)
Active principles	Not known
Other chemical components	Not known
Products in which used	Baked goods, frozen dairy, soft candy, gelatin, puddings, non-alcoholic beverages, alcoholic beverages, chewing gum (3)
Level of use	Cananga oil: baked goods 28.51 ppm, frozen dairy 10.30 ppm, soft candy 31.23 ppm, gelatin, puddings 32.31 ppm, non-alcoholic beverages 13.30 ppm, alcoholic beverages 28.00 ppm, chewing gum 0.36 ppm (4)
Preparation	Essential oil
Main toxicological data	No relevant data found
Data needed	No data required
Specific observations	None
Classification and limits	Herb and essential oil: category 2

National/int. evaluation Cananga oil: Specifications in Food Chemical Codex
USA: Cananga oil: CFR 182.20

Main references (1) Flavour Fragr. J., 11, p.333, (1996)
(2) J. Essent. Oil Res., 9, p.67, (1997)
(3) Leung, 2nd Edition, p.115 (1996)
(4) Fenaroli, 1, p.63, (1995)

Data bases used Chemical Abstracts, Toxline – 1997
Keywords: Cananga

Carica papaya L.

CE No.	109
Steinmetz No.	246
FEMA No.	
Order	Violales
Family	Caricaceae
Name	E Papaw tree, papaya, melon tree, mamao, fructa bomba, lechosa, melon, zapote
	F Papayer
	D Melonenbaum, Papaya
	I Papavero
	SP Papaya
Synonyms	
Parts used	Fruit (1)
Important constituents	Alkaloid (carpaine, 0,4% in leaves), cyanogenic glucosides (tetraphyllin B and prunasin in leaves), carotenoids, esters, alcohols, carbonyls, oxygenated terpenoids (fruit), limonene 100-500 µg/kg in fruit (2,3,4)
Active principles	Not known
Other chemical components	Not known
Products in which used	Fruit as a foodstuff (1), (4), (5), in jellies, desserts, candies (5), juice as beverage (2), leaves and other parts as vegetable (2)
Level of use	Not known
Preparation	Juice, extract (1). Latex from fruit used to make papain (not a flavouring use (5))
Main toxicological data	One case of immediate hypersensitivity (6). Extracts of latex of unripe fruit exert anti-fertility, anti-implantation activity around 500 mg/kg bw level in rats (7, 8). Unripe fruit has abortifacient effect (9). Low dose 0.1 mg/day) of seed extract caused temporary male rat infertility (10). Chymopapain extracted from latex LD50 mice 82 mg/kg and 92 mg/kg rats (11). δ-limonene: ADI not specified (12)
Data needed	No data required
Specific observations	Alkaloid and cyanogenic glucosides found in leaves but no evidence that it is present in the fruit
Classification and limits	**Fruit: category 1**

National/int. evaluation	UK FACC (1976) Appendix I. Enzyme preparations papain and chymopapain are considered acceptable for use in the food industry in the UK (FAC/REP/35)
Main references	(1) MAFF (1995) (2) Duke (1985) (3) Spencer et al. Am J Bot 71, 10 (1984) (4) Idstein et al. Lebens-Wiss. Technol. 18, 3 (1985) (5) Usher (1974) (6) Ezeoke Afr J Med Sci 14 (1985) (7) Garg, S K Planta Medica 26 (1974) (8) Garg, SK et al. Indian J Med Rs 59 (2)(1970) (9) Gopalakrishnan M et al Indian J Physiol Pharmacol (India) 22 1 (1978) (10) Chinoy N J, George, S,M Acta Europaea Fertil. 14(6) (1983) (11) Simmons, J W et al., Drug Chem Tox 7(3)(1984) (12) JECFA 41st meeting (1993)
Data bases used	Biosis (1973-90) Chemical Abstracts (1967-90) FSTA (1969-90) Toxline, Toxlit, Toxnet (1981-90) Embase (1974-90). *Keywords:* Carica papaya, Pawpaw, Papaw, Papaya.

Carum carvi L.

CE No.	112
Steinmetz No.	250
FEMA No.	
Order	Umbelliflorae
Family	Umbelliferae
Name	E Caraway
	F Carvi, cumin (hollandais), cumin des prés
	D Kümmel
	I Carvi
	SP Alcaravea, extractos
Synonyms	
Parts used	Fruit =seed
Important constituents	Essential oil from seed (yield 3-6%): terpenic hydrocarbons [δ-limonene 33.82-46.00%; myrcene 0.21-0.33%; B-caryophyllene 0.07-0.23%]; terpenic ketones and alcohols [δ-carvone 49.05-62.31%; *trans*-dihydrocarvone 0.13-0.51%; *cis*-dihydrocarvone 0.3-0.19%; dihydrocarceol 0.20-0.23%; *trans*-carveol 0.15-0.42%; *cis*-carveol 0.05-0.18%] (1)
Active principles	Not known
Other chemical components	Not known
Products in which used	Baked goods, meat products, processed vegetables alcoholic and non-alcoholic beverages, frozen dairy, candy, gelatin, puddings, soup, snack foods, condiment, relish (2)
Level of use	Caraway: baked goods 8355 ppm; fats, oils 11 ppm; meat products 1442 ppm; processed vegetables 4000 ppm; condiment, relish 435.6 ppm; soups 532 ppm; snack foods 139 ppm; alcoholic beverages 323.7 ppm; gravies 13.5 ppm (2) Caraway oil: baked goods 225.4 ppm; cheese 0.12 ppm; frozen dairy 43.43 ppm; meat products 146 ppm; condiment, relish 41.71 ppm; soft candy 87.9 ppm; gelatin, puddings 128.1 ppm; soups 0.6 ppm; non-alcoholic beverages 35.09 ppm; alcoholic beverages 142.4 ppm; hard candy 50 ppm; chewing gum 0.45 ppm (2)
Preparation	Infusion (3%), decoction (5%), alcoholic distillate in 75% alcohol and distillation waters, essential oil, oleoresin (from seed) (2)
Main toxicological data	Limonene: ADI not specified (3) Carvone: ADI 0-1 mg/kg bw/d (4)

Data needed	No data required
Specific observations	None
Classification and limits	**Fruit and preparations: category 1**
National/int. evaluation	None
Main references	(1) Perfumer&Flavorist 21, p.62 (1996) (2) Fenaroli 1, p.67 (1995) (3) JECFA 41st meeting (1993) (4) JECFA 37th meeting (1990)
Data bases used	Chemical Abstracts (1965-96)

Castor fiber L.

CE No.	3002 (554)
Steinmetz No.	
FEMA No.	Castoreum extract: 2261; Castoreum liquid: 2262
Order	Rodentia
Family	Castoridae
Name	E Siberian or European beaver
	F Castor
	D Biber
	I -
	SP Castoreo, extractos
Synonyms	Canadian beaver (Castor canadensis Kuhl.) is a different species

Parts used	Scent gland secretion from castor sacs (1) called castoreum (2). Castoreum glands (dried and ground)(3).
Important constituents	In castoreum: castorin (0.33-2.5%), volatile oil (1-2%), benzoic acid, salicylic acid, cinnamic acid, phenols, chavicol, betuligenol, ketones, ionone derivative, castoramines, quinolizine alkaloid. (2,4)
Active principles	Not known
Other chemical components	Not known
Products in which used	Extracts used as flavour components (particularly in vanilla flavourings) in most food and beverage categories (2)
Level of use	Average maximum use usually below 94 ppm (2). Use levels in beverages (1-90ppm), in food (1-40ppm). Castoreum extract: baked goods 68.47 ppm; frozen dairy 26.26 ppm; meat products 2 ppm; soft candy 44.10 ppm; gelatin, puddings 47.34 ppm; non-alcoholic beverages 29.77 ppm; alcoholic beverages 93.69 ppm; gravies 0.60 ppm; hard candy 24.17 ppm; chewing gum 42.09 ppm (3). Castoreum liquid: baked goods 4.87 ppm; frozen dairy 1.46 ppm; soft candy 2.92 ppm; gelatin, puddings 1.25 ppm; non-alcoholic beverages 1.72 ppm; hard candy 9.09 ppm; chewing gum 0.01 ppm (3)
Preparation	Dried secretion, castoreum, extract, absolute, tincture (2). Resinoid prepared by extraction of dried, ground pouches using petroleum ether. Absolute prepared by alcoholic extraction of resinoid (3)
Main toxicological data	Castoreum tincture applied full strength to mice or rabbit skins was not irritating. No irritation or sensitisation effects seen in humans when tested at a concentration of 4% in petrolatum (5). Neither is castoreum tincture phototoxic (5). Reported to have sedative,

	nervine and other properties (2). Cinnamic acid category A Bleu Book 4th Ed. Volume I
Data needed	Quantitative information on chemical components of preparation and if necessary, mutagenicity study and 28-day oral study on the appropriate preparation
Specific observations	
Classification and limits	Castoreum: category 5
National/int. evaluation	UK FACC (1976) Appendix 2 USA GRAS (182.50) FEMA 2261 (Castoreum extract) FEMA 2262 (Castoreum liquid)
Main references	(1) Chem Scr. 24, 2, p100-103 (1984) (2) Leung (1996) (3) Fenaroli (1995) (4) Helv. Chim. Acta, 59, 4, p1169-85 (1976) (5) Food Cosmet. Toxicol 11, 1061 (1973)
Data bases used	Chemical Abstracts (1967-97) Biosis (1973-88) FSTA (1969-88) Medline (1966-97) Embase (1982-88) CAB (1970-88) Toxline (1981-97) *Keywords:* Castor fiber or canadensis, Castoreum, Beaver. Castor, Castoramine, Castorin

Centaurium erythraea Rafn.

CE No.	183
Steinmetz No.	
FEMA No.	
Order	Gentianales
Family	Gentianaceae
Name	**E** Minor centaury, common centaury, pink centaury
	D Echtes Tausendgüldenkraut
	F Petite centaurée, erythrée, fiel de terre, herbe à la fièvre, herbe à mille-florins
	I Centaurea minore, biondella, caccia febbre, fiel de terra
	SP
Synonyms	C. erythraea Rafn. ssp. erythraea, C. minus auct. non Moench, Erythraea centaurium auct. non (L.) Pers., C. umbellatum Gilib.

Parts used	Flowering tops (1)
Important constituents	The dried aerial parts contain small amounts of the intensively bitter-tasting secoiridoidglycosides: swertiamarin (2-8%, 75% of the total iridoid content), sweroside (0.1-0.5%) and gentiopicrin (syn. gentiopicroside, 0.1-0.5%). Bitter taste mainly due to the small amounts of m-hydroxybenzoic acid esters of sweroside, centapicrin and desacetylcentapicrin (0.1-0.2% in capsules). Of the same chemical group are decentapicrin A and B (2,3). Furthermore another secoiridoid gentioflavosid (4), and the iridoids dihydrocornin (4) and 6'-(m-hydroxybenzoyl)loganin (4,5), also xanthon X2 (up to 0.2%) (4). Small amounts of a dimeric secoiridoid, centaurosid (4,5). Furthermore 0.2% flavonoids (6). Various polyoxygenated xanthones (ca. 0.05%): methylbellidifolin (0.004% in roots), 8-desmethyleustomin (0.009% in roots), decussatin (0.002% in roots), eustomin (0.02% in roots) and others mainly in the roots, the two dihydroxylated xanthones methylbellidifolin and 8-desmethyleustomin also present in aerial parts (6), triterpenes, sterols (steroidic fraction with 5.6% brassicasterol, 14.8% campesterol, 34.0% stigmasterol, 41.1% b-sitosterol and 4.5% Δ7-stigmastenol) (7), oleanolic acid (1) present (5). Reports on the presence of traces (0.3%) of pyridine- and actinidin alkaloids (secoiridoidalkaloids gentianin and gentianidin) are controversial and most probably due to isolation artefacts (5,8). Rich in phenolic acids with a C_6-C_1 or a C_6-C_3 unit (0.27% of dry wt; i.e. protocatechic, meta and parahydroxybenzoic, vanillic, syringic, paracoumaric, cafeic, ferulic and sinapic acid, also 3,4-dihydrophenylacetic acid) (9)

	In contrast to earlier reports no amarogentin present (5). Erythrocentaurin possibly also present (1)
Active principles	Xanthones
Other chemical components	Not known
Products in which used	Used for the formulation of bitter tonics. The dried product, the tincture, and the fluid extract blend very well with other bitter flavours, because, while bitter, they contribute little other aromatic flavour (1)
Level of use	Tincture of the herb used in non-alcoholic beverages 115 ppm and alcoholic beverages 4238 ppm; tincture of the flowers used in non-alcoholic beverages 30 ppm and alcoholic beverages 700 ppm (IOFI 1997)
Preparation	Tincture (20% in 20-40% ethanol), fluid extract (1), water infusion (10)
Main toxicological data	No reported side-effects or documented toxicity data were located for centaury (10) Genotoxicity: Negative Ames test for the xanthones dimethylbellidifolin (1,3,5-methoxylated), bellidifolin (3-methoxylated) and desmethylbellidifolin (not methoxylated) with S. typhimurium TA98 and TA100 (positive with TA97 and TA2637) with and without metabolic activation. Positive Ames test for 7-methoxylated xanthones with metabolic activation (11)
Data needed	No data required
Specific observations	None
Classification and limits	**Flowering tops: category 4 (with limits on xanthones)**
National/int. evaluation	None
Main references	(1) Fenaroli (1995) (2) Planta medica 41: 221-231 (1981) (3) Planta medica 41: 150-160 (1981) (4) Haenseler, R. et al. Hager's Handbuch der Pharmazeutischen Praxis. 5th Ed., Springer Verlag, Berlin (1990) (5) DAB 9. Deutsches Arzneibuch – Kommentar. 9th Ed., WVG, Stuttgart (1987) (6) J. Nat. Prod. 49 (2): 359-361 (1986) (7) Boll. Soc. It. Biol.Sper. 61 (2): 165-169 (1985) (8) Planta medica 33: 422-423 (1978) (9) Ann. Pharm. Franc. 35 (3-4): 107-111 (1977) (10) Newall, C.A. et al. Herbal Medicines – A guide for health-care professionals. The Pharmaceutical Press, London (1996) (11) Mutat. Res. 150: 141-146 (1985)
Data bases used	Medline (1966-97) Embase (1980-97)

Toxline (1965-97)
Biological Abstracts (1989-97)
Keywords: Centaurium erythraea, Centaurium umbellatum, Erythraea centaurium, minor centaury

Cinchona officinalis L.

CE No.	2027
Steinmetz No.	
FEMA No.	
Order	Gentianales
Family	Rubiaceae
Name	E Yellow cinchona, ledger cinchona
	F Quinquina jaune royal
	D Gelbe Chinarinde, Königschinarinde
	I China
	SP
Synonyms	C. calisaya Wedd., C. calisaya Wedd. var. ledgeriana Howard, C. ledgeriana Moens ex Trim (according to the newer botanical nomenclature there are two types distinguished: C. officinalis type C. calisaya and C. officinalis type C. ledgeriana)
Parts used	Bark of the branches (1)
Important constituents	See Cinchona pubescens
	The bark of C. officinalis has a higher content in quinine and a lower content in other alkaloids than C. pubescens (2). Cortex Chinae calisaya: Total content in alkaloids of 4-8% with over 50% quinine alkaloids in the bark. Cortex Chinae ledgerianae: Total content in alkaloids of up to 15% with over 80% to 90% quinine alkaloids and a maximum of 4% quinidine in the bark (2). Highest alkaloid content of all Cinchona ssp. in C. officinalis type C. ledgeriana
Active principles	Not known
Other chemical components	Quinine
Products in which used	Cinchona bark and bark extract are used in non-alcoholic and alcoholic beverages. Cinchona extract (Cinchona ssp.) is used in soft candies, non-alcoholic and alcoholic beverages (1)
Level of use	Bark: non-alcoholic beverages 4.40 ppm, alcoholic beverages 132.3 ppm; bark extract: non-alcoholic beverages 140.7 ppm, alcoholic beverages 500.0 ppm; Cinchona extract (Cinchona ssp.): soft candy 3.00 ppm, non-alcoholic beverages 113.3 ppm, alcoholic beverages 466.9 ppm (1)
Preparation	Tincture (20% in 70% ethanol or 10% in 60% ethanol both tannin-free or not); fluid, soft, and dried extracts (the last two can be aqueous or water-alcoholic extracts with or without tannin) (1)
Main toxicological data	See CE No.128

Data needed	No data required
Specific observations	None
Classification and limits	**Bark: category 4 (with limits on quinine)**
National/int. evaluation	Regulatory status in USA: CFR 172.510 (A.B.) in beverages only (not more than 83 ppm total cinchona alkaloids in finished beverages), FEMA No. 2283 (Cinchona bark, yellow), 2284 (Cinchona bark, yellow, extract), 2285 (Cinchona extract) (1)
Main references	(1) Fenaroli (1995) (2) Hager's Handbuch der Pharmazeutischen Praxis, 5th Ed., Haenseler R. et al. (Ed.), Springer Verlag, Berlin (1990)
Data bases used	Medline (1966-98) Embase (1980-98) Biological Abstracts (1989-98) CC Life (2/97-2/98) *Keywords:* Cinchona calisaya, Cinchona ledgeriana

Cinchona pubescens Vahl

CE No.	128
Steinmetz No.	-
FEMA No.	Cinchona bark, red: 2281, Cinchona bark, red extract: 2282, Cinchona extract: 2285
Order	Gentianales
Family	Rubiaceae
Name	E Red cinchona, Jesuit's bark, Peruvian bark
	F Quina, Quinquina rouge
	D Rote Chinarinde, Fieberrinde
	I Quina
	SP Quina
Synonyms	Cinchona succirubra Pav. ex Klotsch, C. cordifolia Mut. ex. Humb., C. tucuyensis Karst
Parts used	Bark of the branches (1)
Important constituents	The bark contains 5-15% total quinoline alkaloids (average 6-10% of which 30-60% are quininetype alkaloids) that consist mainly of quinine (1-3%;), quinidine (0-4%), and their respective 6'-demethoxy derivatives cinchonidine (1.25-8%) and cinchonine (2-8%). Minor amounts of other alkaloids such as dihydroquinine, dihydroquinidine, dihydrocinchonidine, dihydrocinchonine, quinamine, epiquinamine, epiquinine, hydroquinine and others. Content of total alkaloids lower than in C. officinalis type ledgeriana. Other constituents include norsolorinic acid (0.0008%, an anthraquinone), tannins (cinchotannic acid and its decomposition product 'cinchona red' (a phlobaphene), cinchonains (precursors of cinchona bark tannins), bitter glycosides (quinovic acid esters), acids (5-8% quinic acid, caffeic acid), b-sitosterol, starch, resin, wax, traces (0.005%) of essential oil (2,3,4,5,6)
Active principles	Not known
Other chemical components	Quinine
Products in which used	Cinchona bark is used in baked goods, the bark extract is used in baked goods, frozen dairy, condiment/relish, soft candy, non-alcoholic (tonic water) and alcoholic beverages (bitters and liqueurs). Cinchona extract (Cinchona ssp.) is used in soft candies, non-alcoholic and alcoholic beverages (1)
Level of use	Bark: baked goods 27 ppm; *bark extract:* baked goods 16.3 ppm, frozen dairy 40.00 ppm, condiment/relish 85.00 ppm, soft candy 20.00 ppm, non-alcoholic beverages 146.2 ppm, alcoholic bever-

ages 278.3 ppm; *Cinchona extract* (Cinchona ssp.): soft candy 3.00 ppm, non-alcoholic beverages 113.3 ppm, alcoholic beverages 466.9 ppm (1)

Preparation

Tincture (20% in 70% ethanol or 10% in 60% ethanol both tannin-free or not); fluid, soft, and dried extracts (the last two can be aqueous or water-alcoholic extracts with or without tannin) (1)

Main toxicological data

Of the various alkaloids (about 35 indol-, bis-indol- and quinoline alkaloids) of cinchona bark, quinine is the most important with respect to toxicology and therapeutic use. Quinidine is the dextrorotatory stereoisomer of quinine, and has qualitatively the same pharmacological effects. Quinine has antipyretic effects. It is locally anesthetizing and muscle-relaxing due to a reduced reaction on nervous stimulus and acetylcholine concentrations at the muscle end-plates. Quinine is therapeutically used at oral doses of 650 mg quinine sulfate three times a day for 10-14 days to treat malaria due to its activity against the schizontic form of Plasmodium falciparum. It is also cytotoxic to various procaryontes (bacteria, yeasts, trypanosomes, ciliates). Quinine was shown to be a strong monoamino oxidase inhibitor in vitro, while cinchonine and cinchonidine were less potent (7). Quinidine is mainly used for treating cardiac arrhythmias (class Ia of antiarrhytmicas according to Vaughan Williams; quinidine sulfate p.o. 3-4x 200-300mg/d, daily dose of 600-1200mg) as it blocks the sodium influx and the potassium efflux at the heart muscle cells. It has a lower anticholinergic activity than quinine.

Cinchona alkaloids have a toxic potency. Intoxication (cinchonism) is usually due to hypersensitivity or overdosage (5). Ground cinchona bark and quinine alkaloids have been reported to cause skin reactions such as urticaria, contact dermatitis, and other hypersensitive reactions in humans (2,8). Thrombocytopenia resulting in bleeding has been observed in rare cases. Thus, interactions may occur with anticoagulans (8). Unwanted side effects of therapeutic treatment with quinine occur quite often and result in skin reactions with erythema, exanthema, increased perspiration, and sometimes oedema. Apart from serious central nervous disturbances, cardiac complications and irreversible nerve damage may also occur (9). Unwanted effects at therapeutic doses of quinidine are rarely occurring rheumatoid effects as a result of allergic reactions, and at the heart, proarrhythmic effects with tachycardia and ventricular fibrillation.

Unwanted effects with overdose or chronic use of quinine alkaloids are nausea, vomiting, diarrhea and abdominal pain, severe headache, disturbed vision, delirium, convulsions, paralysis, and collapse. Very high doses may lead to visual and auditory defects (8). Cinchona bark and quinine have embryotoxic effects and are abortive at overdosage. No teratogenic effects, however, have been observed in chronic feeding studies with rats at concentrations of up to 200 mg/kg bw quinine hydrochloride. There were no indications of interference with auditory function in a 13-week study in rats receiving up to 200 mg/kg bw quinine hydrochloride (10). Serious intoxications by quinine may occur in children, adults with heart disease and healthy adults after oral doses of 1-2 g, 2 g and 6 g quinine, resp. (5). A single oral dose of 8 g quinine may be fatal to an adult (8,9). No unwanted effects have been observed in humans after a daily uptake of 100 to 120 mg quinine chloride in beverages for 14 consecutive days (11). A daily dose of 40 mg quinine chloride (in 1 litre of tonic water) has been regarded as tolerable to healthy adults (10)

Data needed	No data required
Specific observations	None
Classification and limits	Bark: category 4 (with limits on quinine)
National/int. evaluation	Regulatory status in USA: CFR 172.510 (A.B.) in beverages only (not more than 83 ppm (0.0083%) total cinchona alkaloids in finished beverages (1)
Main references	(1) Fenaroli (1995) (2) Leung (1996) (3) Wichtl M., Teedrogen. 2nd Ed., WVG, Stuttgart (1989) (4) Hoppe H.A., Drogenkunde, 8th Ed., de Gruyter, Berlin (1975) (5) Hager's Handbuch der Pharmazeutischen Praxis, 5th Ed., Haenseler R. et al. (Ed.), Springer Verlag, Berlin (1990) (6) Trease W.C., Trease and Evans' Pharmacognosy, 14th Ed., WB Saunders, London (1996) (7) Chem. Pharm. Bull. 37: 363-366 (1989) (8) Martindale, The Extra Pharmacopoeia, 29th Ed., Pharmaceutical Press, London (1989) (9) Wirth W., Gloxhuber Ch., Toxikologie, 5th Ed., Thieme, Stuttgart (1994) (10) Toxicology 54: 219-226 (1989) (11) Lancet Jan 31: 271-272 (1987)
Data bases used	Medline (1966-98) Embase (1980-98) Biological Abstracts (1989-98) CC Life (2/97-2/98) *Keywords:* Cinchona calisaya, Cinchona ledgeriana

Cistus ladanifer L.

CE No.	134
Steinmetz No.	309
FEMA No.	Absolute: 2608; Oil: 2609; Oleoresin: 2610
Order	Violales
Family	Cistaceae
Name	**E** Rock rose
	F Ladanum d'Espagne
	D Lack-Cistrose
	I Labdano
	SP Labdano
Synonyms	C. viscosus Stokes C. grandiflorus Pourr., C. landanosma Hoffm. C. ladanifer L., Landanum officianarum Spach
Parts used	Resin gum (labdanum) from leaves and twigs
Important constituents	Essential oil (1, 2, 3): α-pinene (3.5%), camphene (11%), limonene (2.6%), γ-terpinene (2%), bornyl acetate (4.2%), α-thujone (0.8%), carophyllene (0.47%), δ-phellandrene (0.41%), linalool (0.5%), eucalyptol (0.2%),?-cadinene,(0.3%) 2,2,6-trimethyl-cyclohexa-none (5.7%), benzaldehyde, myrcene, p-cymene (4%), eugenol (trace). (Levels of some componenents vary depending on the time of year)
	Labdanum resin (4, 5, 6): labdanoic and labdenoic acid derivatives, neutral labdane compounds, kaempferol and apigenin derivatives
Active principles	Thujone
Other chemical components	Eucalyptol
Products in which used	Baked goods, frozen dairy, soft candy, gelatin, puddings, non-alcoholic beverages, alcoholic beverages, chewing gum, meat products, condiment relish (7)
Level of use	Labdanum absolute: baked goods 20.33 ppm; frozen dairy 9.58 ppm; soft candy 12.07 ppm; gelatin, puddings 8.49 ppm; non-alcoholic beverages 3.88 ppm; alcoholic beverages 4.07 ppm; chewing gum 0.10 ppm
	Labdanum oil: baked goods 2.40 ppm; frozen dairy 2.05 ppm; meat products 2.00 ppm; condiment, relish 2.00 ppm; soft candy 2.20 ppm; sweet sauce 3.50 ppm; gelatin, puddings 2.02 ppm; non-alcoholic beverages 1.80 ppm; alcoholic beverages 1.86 ppm; hard candy 0.08 ppm; chewing gum 0.47 ppm
	Labdanum oleoresin: baked goods 14.11 ppm; frozen dairy 6.79 ppm; soft candy 9.70 ppm; gelatin, puddings 8.00 ppm; non-alco-

holic beverages 4.56 ppm; alcoholic beverages 4.00 ppm(7)
NB: figures could include preparations from other Cistus spp. although Food Chemicals Codex (8) defines Labdanum oil as being derived from the gum of C. ladaniferus L. only. C. ladaniferus L. resin and gum absolute in beverages (0.5-10ppm), in food (0.2-23 ppm)

Preparation Oleoresin (labdanum gum); absolute from leaves and twigs. Essential oil from (a) gum; (b) leaves and twigs directly (Cistus oil) (9, 10)

Main toxicological data Cistus ladaniferus extract (5 x 1.6mg ip) causes convulsions in mice (11) Hydro-alcoholic extract (1.81g/kg/day for 26 days) causes lesions in lung, liver and kidneys in rats (12). Thujone under evaluation as an active principle. Kaempferol is mutagenic but little evidence of carcinogenicity. Most essential oil constituents category B Blue Book 4th Ed. Volume I. Linalool, benzaldehyde category A. D-Limonene: ADI not specified (13)

Data needed Clarification of use of Cistus oil/Labdanum oil. Data on chemical components of labdanum oil and absolute, and if necessary 28-day oral study and mutagenicity study. Toxicity data on hydro-alcoholic extract

Specific observations None

Classification and limits **Resin gum: category 5 (with limits on eucalyptol and thujone)**
Herb: category 5 (with limits on eucalyptol and thujone)

National/int. evaluation UK FACC Report (1976) Cistus ladaniferus resin gum Appendix 2
USA- absolute, oil and oleoresin approved for food use (172.510)
FEMA: 2608 absolute; 2609 oil; 2610 oleoresin
Thujone use restricted as in Annex II of Council Directive 88/388/EEC

Main references
(1) Z. Naturforsch 39c 699, (1984)
(2) Anales de Quimica 83(2): 201, (1987)
(3) Z Pflanzenphysiol 72: 237 (1974)
(4) Perf.&Flavorist 9(1): 49 (1984)
(5) Z. Naturforsch 39: 303 (1984)
(6) Phytochem. 23(2): 470-1 (1984)
(7) Fenaroli (1995)
(8) Food Chem. Codex (1981)
(9) Perfums. Cosmet. Aromes 67:59 (1986)
(10) Arctander (1960) Perfum. & Flavour Matter of Nat Origin.
(11) Revue Med. Vet 138(12):971 (1987)
(12) Medicina Veterinaria 3(4):219 (1986)
(13) JECFA 41st meeting (1993)

Data bases used Chem. Abs. (1967-97)
Biosis(1973-89)
FSTA(1969-89)
CAB(1970-89)

Medline (1966-89)
Embase (1982-89)
Keywords: Cistus ladaniferus, C. Creticus C. polymorphus, Labdanum

Citrus aurantiifolia (Christm.) Swingle

CE No.	141
Steinmetz No.	316
FEMA No.	Essential oil, expressed: 2631
Order	Rutales
Family	Rutaceae
Name	E Lime tree
	F Limier
	D Sauerlimettenbaum, Limettenbaum
	I Limetta
	SP Lima, aceite esencial
Synonyms	C. lima Lunan; C. medica ssp. acida (Roxb.) Engl.
Parts used	Fruit, pulp, rind, leaf and small twigs
Important constituents	Rind expressed (or cold pressed) essential oil: terpenic hydrocarbons [limonene 34-64%; γ-terpinene 7-21%; β-bisabolene 2-5.7%; myrcene; α- and β-pinene; α-bergamottene; p.cymene]; aldehydes [citral 1-6%; decanal]; terpenic alcohols [α-terpineol; linalool 0.1-0.3%; eucalyptol] (1,2); furocoumarins: limettin, bergapten (3) Rind distilled essential oil: terpenic hydrocarbons [limonene, α- and β-pinene; γ-terpinene; myrcene, p.cymene]; citral; α-terpineol; eucalyptol; geranyl acetate Lime leaf oil (produced in Japan): limonene 24.7%; α-pinene 0.3%; α-thujene 0.2%; β-pinene 0.3%; sabinene 2.3%; myrcene 1.0%; δ-3-carene 0.4%; (E)-b-ocimene 1.4%; γ-terpinene 0.3%; citronellal 0.55; decanal 0.2%; neral 23.8%; geranial 40%; linalool 0.8%; nerol 0.2%; geraniol 0.6%; a-terpineol 0.2%; geranyl acetate 0.5% (4)
Active principles	Furocoumarins
Other chemical components	Eucalyptol
Products in which used	Non-alcoholic beverages, alcoholic beverages, frozen dairy, desserts, baked goods, gelatin, puddings, meat and meat products, candy
Level of use	Lime oil: Baked goods 557.4 ppm; breakfast cereals 199.7 ppm; meat products 84.01 ppm; condiment relish 100 ppm; soft candy 782.9 ppm; gelatin, puddings 477.3 ppm; non-alcoholic beverages 75.54 pppm; alcoholic beverages 452.1 ppm; gravies 1.38 ppm; hard candy 948.2 ppm; chewing gum 2322 ppm (5) Lime terpeneless oil: Baked goods 100.1 ppm; frozen dairy 83.80 ppm; meat products 10 ppm; condiment, relish 10 ppm; soft candy 87.71 ppm; gelatin, puddings 91.97 ppm; non-alcoholic beverages

	45.1 ppm; alcoholic beverages 122.3 ppm; hard candy 52.64 ppm; chewing gum 3.77 ppm (5)
Preparation	Essential oil, terpeneless essential oil, soluble oil
Main toxicological data	Phototoxicity of expressed lime oil (6) δ-limonene: ADI not specified (7) Citral (neral + geranial), citronellal, linalool, geranyl acetate: ADI 0-0.5 mg/kg bodyweight and day (8) Furocoumarins: under evalutation
Data needed	No data required
Specific observations	None
Classification and limits	**Fruit: category 1** **Rind: category 1** **Lime distilled oil: category 3 (with limits on eucalyptol and furocoumarins)** **Lime expressed oil: category 3 (with limits on eucalyptol and furocoumarins)** **Lime leaf oil: category 1**
National/int. evaluation	CFR chap. 182.20, 580.20 FACC 1976 Appendix 2
Main references	(1) CIVO-TNO: Volatile compounds in Food Ed. VI:87-92 (2) Lawrence, Essential oils (1976-91) (3) Perfumer&Flavorist 7(3):57 (1982) (4) Perfumer&Flavorist 18:54 (1993) (5) Fenaroli Vol 1, p. 175 (1995) (6) Food Cosmet. Toxicol. 12:729 (1974) (7) JECFA 41st meeting (1993) (8) JECFA 23rd session (1979)
Data bases used	Chemical abstracts 1956-96 *Keywords:* Lime oil

Citrus aurantium L. ssp. aurantium L.

CE No.	136 a
Steinmetz No.	312
FEMA No.	Rind oil: 2823
Order	Rutales
Family	Rutaceae
Name	E Neroli, bitter orange tree
	F Oranger amer, bigaradier
	D Bitterorangenbaum, Pomeranzenbaum
	I Arancio amaro neroli
	SP Naranjo amargo
Synonyms	C. aurantium L. ssp. amara Engl..
Parts used	Fruit: pulp, 136a rind , (136b flower , 136c leaf and twig)
Important constituents	Rind expressed (or cold pressed) essential oil: terpenic hydrocarbons [mainly limonene ~ 90%, myrcene, $\alpha+\beta$-pinene]; terpenic alcohols and esters [linalool, α-terpineol, nerolidol, linalyl acetate]; aldehydes [decanal, nonanal]; trace amounts of furocoumarins (ref 1,2)
	Rind infusion, tincture: qualitative composition: essential oil + bitter compounds (flavonoids)
Active principles	Furocoumarins
Other chemical components	Not known
Products in which used	Alcoholic and non-alcoholic beverages, frozen dairy, desserts, candy, baked goods, gelatin and puddings
Level of use	(IOFI, 1997):
	Rind expressed oil: non-alcoholic beverages 80 ppm, alcoholic beverages 970 ppm, ices 109 ppm, candy 315 ppm, baked goods 588 ppm, gelatin, desserts 320 ppm, others 1030 ppm
	Rind expressed terpeneless oil: non-alcoholic beverages 25 ppm, alcoholic beverages 26 ppm, ices 50 ppm, candy 100 ppm, baked goods 500 ppm, gelatin, desserts 50 ppm, others 2000 ppm.
	Rind essential oil: non-alcoholic beverages 160 ppm, alcoholic beverages 660 ppm, ices 490 ppm, candy 750 ppm, baked goods 1800 ppm, gelatin, desserts 800 ppm, others 10000 ppm.
	Rind essential terpeneless oil: non-alcoholic beverages 25 ppm, alcoholic beverages 45 ppm, ices 50 ppm, candy 100 ppm, baked goods 500 ppm, gelatin, desserts 50 ppm , others 2000 ppm.
Preparation	Juice, essential oil, deterpenated essential oil, infusion, tincture

Main toxicological data	Phototoxicity on Bitter Orange oil (ref 3) d-Limonene: ADI not specified (ref 4) Linalool and linalyl acetate: ADI: 0,5 mg/kg bw/d (ref 5) Furocoumarins: under evaluation
Data needed	No data required
Specific observations	Fruit is used as a foodstuff
Classification and limits	**Rind, infusion: category 3 (with limits on furocoumarins)** **Essential oil: category 3 (with limits on furocoumarins)**
National/int. evaluation	Neroli bigarde oil: FDA 182.20
Main references	(1) CIVO-TNO "Volatile compounds in Food" Ed. VI, p. 50-51 (2) Perf. Flav., 18, p.43, (Sept/Oct 1993) (3) Food Cosmet. Toxicol., 12, p.735, (1974) (4) JECFA 41st meeting, 1993 (5) JECFA 23rd session, 1979
Data bases used	Chemical Abstracts: 1956-96 *Keywords:* Bitter orange oil

Citrus aurantium L. ssp. aurantium L.

CE No.	136 b
Steinmetz No.	312
FEMA No.	Neroli bigarade oil: 2771; Orange flower absolute: 2818; Orange flowers: 2819
Order	Rutales
Family	Rutaceae
Name	E Neroli, bitter orange tree F Oranger amer, bigaradier D Bitterorangenbaum, Pomeranzenbaum I Arancio amaro, neroli SP Naranjo amargo
Synonyms	C. aurantium L. ssp. amara Engl.
Parts used	Fruit: pulp, (136a rind), 136b flower, (136c leaf and twig)
Important constituents	Neroli essential oil (bitter orange flower oil): (yield: ~ 0,1%) terpenic hydrocarbons [mainly limonene 9-18%, α-pinene < 2%, β-pinene 7-17%, myrcene]; terpenic alcohols and esters: [linalool 28-44%, α-terpineol 2-5.5%, nerolidol 1-5%, farnesol 1-4%, linalyl acetate 3-15%, neryl acetate < 2.5%, geranyl acetate 1-5%] (1,2)
Active principles	Not known
Other chemical components	Not known
Products in which used	Alcoholic and non-alcoholic beverages, frozen dairy, desserts, candy, baked goods, gelatin and puddings
Level of use	(IOFI, 1997): Neroli essential oil: non-alcoholic beverages 25 ppm, alcoholic beverages 7 ppm, ices 15 ppm, candy 26 ppm, baked goods 35 ppm, gelatin, desserts 30 ppm, others 75 ppm Bitter orange flowers: 100-2000 ppm Bitter orange blossom absolute: non-alcoholic beverages 6 ppm, alcoholic beverages 14 ppm, ices 13 ppm, candy 51 ppm, baked goods 51 ppm, gelatin, desserts 12 ppm, others 52 ppm Bitter orange concrete: non-alcoholic beverages 1.5 ppm, alcoholic beverages 0.5 ppm, ices 0.3 ppm, candy 0.65 ppm, gelatin, desserts 0.5 ppm.
Preparation	Essential oil, absolute, concrete, distilled water, distilled water extract.
Main toxicological data	Linalool; linalyl acetate; geranyl acetate: ADI: 0,5 mg/kg bw/d (3) d-Lmonene: ADI not specified (4)

	Geranyl acetate: not carcinogenic for F344/N rats or B6C3F1 mice (5)
Data needed	No data required
Specific observations	Bitter orange flower distilled water is used as a foodstuff
Classification and limits	**Flower, distilled water: category 1** **Essential oil, absolute: category 1**
National/int. evaluation	Neroli oil and Orange flower absolute: FDA 182.20, 582.20
Main references	(1) Perf. Flav., 9, p.7, (Feb/Mar 1985) (2) Leung, p. 250, (1980) (3) JECFA 23rd Session, 1979 (4) JECFA 41st Session, 1993 (5) NTP Technical Report No. 252, October 1987
Data bases used	Chemical Abstracts: 1956-96 *Keywords:* Neroli oil – Bitter orange flower oil

Citrus aurantium L. ssp. aurantium L.

CE No.	136 c
Steinmetz No.	312
FEMA No.	Petitgrain bigarade oil: 2855
Order	Rutales
Family	Rutaceae
Name	E Neroli, bitter orange tree
	F Oranger amer, bigaradier
	D Bitterorangenbaum, Pomeranzenbaum
	I Arancio amaro, neroli
	SP Naranjo amargo
Synonyms	C. aurantium L. ssp. amara Engl.
Parts used	Fruit: pulp, (136a rind, 136b flowers), 136c leaf and twig
Important constituents	Petitgrain oil = Bitter orange leaf and twig oil: (reference samples France): terpenic alcohols and esters [linalool 12.3-24.2%, α-terpineol 2.1-5.0%, geraniol 1.4-2.3%, nerol 0.4-1.1%, linalyl acetate 51.0-71.0%, neryl acetate 1.1-2.0%, geranyl acetate 1.9-3.4%]; terpenic hydrocarbons [limonene 0.4-8%, β-pinene 0.3-2.7%, myrcene 0-1.5% (E)-β-ocimene 0.2-2.2%] (1) Italian bitter orange petitgrain oil: terpenic alcohols and esters [linalool 21.7-32.55%, α-terpineol 3.09-5.63%, geraniol 0.71-0.95%, nerol 0.75-0.99%, terpinen-4-ol 0.05-0.08%, linalyl acetate 50.68-62.57%, neryl acetate 1.4-1.73%, geranyl acetate 1.9-3.16%]; terpenic hydrocarbons [limonene 0.44-2.17%, β-pinene 0.65-1.15%, myrcene 0.56-1.24%, (E)-β-ocimene 0.57-1.76%, (Z)-β-ocimene 0.20-0.44%], + methyl N-methyl anthranilate (traces-1.0%) (2)
Active principles	Not known
Other chemical components	Not known
Products in which used	Non-alcoholic beverages, alcoholic beverages, ices, candy, baked goods, gelatin, desserts
Level of use	Leaf and twig essential oil: non-alcoholic beverages 600 ppm; alcoholic beverages 1000 ppm; ices 1150 ppm; candy 1400 ppm; baked goods 2100 ppm; gelatin, desserts 2100 ppm; others 2400 ppm (IOFI, 1997)
Preparation	Leaf and twig essential oil
Main toxicological data	Linalool and linalyl acetate: ADI: 0.5 mg/kg bw/d (4) Methyl N-methyl anthranilate: ADI: 0.2 mg/kg bw/d (4)

	Geranyl acetate: not carcinogenic in F344 rats and B6C3F1 mice (5)
	d-Limonene: ADI not specified (6)
Data needed	No data required
Specific observations	None
Classification and limits	Bitter orange leaf oil: category 2
National/int. evaluation	Petitgrain bigarade oil: FDA 182.20, 582.20
Main references	(1) Perf. Flav., 18(5), p.43, (1993)
	(2) Perf. Flav., 21(3), p.17, (1996)
	(3) Fenaroli, Vol.1, p.226, (1995)
	(4) JECFA 23rd session, 1979
	(5) NTP Technical Report No.252, October 1987
	(6) JECFA 41st Meeting 1993
Data bases used	Chemical Abstracts: 1956-96
	Keywords: Petitgrain oil

Citrus aurantium L. ssp. bergamia (Risso&Poit.)

CE No.	137
Steinmetz No.	313
FEMA No.	Essential oil of rind: 2153
Order	Rutales
Family	Rutaceae
Name	E Bergamot tree
	F Bergamotier
	D Bergamott-Orangenbaum, Bergamottenbaum
	I Bergamotto
	SP Bergamota
Synonyms	Citrus bergamia (Risso&Poiteau)

Parts used	Fruit, rind, leaf and twig
Important constituents	Rind expressed (or cold pressed) oil (yield 0.5%): terpenic hydrocarbons [limonene 33-42%; γ-terpinene -10%; α-pinene 5-9%; β-pinene]; terpenic alcohol: linalool 6-15%; terpenic ester: linalyl acetate 23-25% (1,2,3,4,5); furocoumarins: bergaptene (2000-4500 ppm); bergamottin (6)
	Leaf essential oil: terpenic alcohols [linalool 22-55%; terpineol 4-5%; terpinene.q.ol.4 1.5-2.5%; nerol; geraniol]; terpenic ester: linalyl acetate 22-52%; terpneic hydrocarbons [myrcene; β-ocimene; limonene] (1,2,3)
Active principles	Furocoumarins
Other chemical components	Not known
Products in which used	Alcoholic and non-alcoholic beverages, frozen dairy desserts, candy, baked goods, gelatin and puddings, meat and meat products
Level of use	Rind essential oil: baked goods 92.97 ppm; frozen dairy 49.47 ppm; meat products 0.15 ppm; soft candy 95.19 ppm; gelatin, puddings 200 ppm; non-alcoholic beverages 68.56 ppm; alcoholic beverages 91.13 ppm; hard candy 1.96 ppm; chewing gum 5.57 ppm (7)
Preparation	Rind essential oil and rectified essential oil; leaf essential oil
Main toxicological data	d-Limonene: ADI not specified (8)
	Linalool and linalyl acetate: ADI 0-0.5 mg/kg bw/d (9)
Data needed	No data required
Specific observations	None

Classification and limits	Fruit: category 1 Rind: category 3 (with limits on furocoumarins) Rind essential oil: category 3 (with limits on furocoumarins) Leaf essential oil: category 2
National/int. evaluation	CFR chap. 182.20 FACC 1976 Appendix 1 (fruit rind)
Main references	(1) Perfumer&Flavorist 18(5):62 (1993) (2) Lawrence, Essential oils (1976-91) (3) AFNOR NFT Sept:75-215 (1989) (4) Perfumer&Flavorist 21(4):25 (1996) (5) Perfumer&Flavorist 21(6):19 (1996) (6) Perfumer&Flavorist 7(3):57 (1982) (7) Fenaroli Vol 1, p 47 (1995) (8) JECFA 41st meeting (1993) (9) JECFA 23rd session (1979)
Data bases used	Chemical Abstracts 1956-96) *Keywords:* Bergamot oil, petitgrain bergamot oil

Citrus aurantium L. var. myrtifolia Ker-Gawl.

CE No.	138
Steinmetz No.	-
FEMA No.	-
Order	Rutales
Family	Rutaceae
Name	E Myrtle-leaved orange tree
	F Oranger à feuilles de myrte
	D Myrtenblättriger Orangenbaum, Pomelle
	I Arancio a foglia mirtella
	SP
Synonyms	Citrus myrtifolia Raf., Citrus auriantum ssp. amara var. pumilia
Parts used	Fruit, rind
Important constituents	Essential oil of dried peel: terpenic hydrocarbons [limonene 80%; α- and β-pinene; myrcene; b-ocimene]; terpenic alcohols [linalool 5.5%; α-terpenieol; terpinen-4-ol; geranil; nerol; farnesol; nerolidol]; linalyl acetate 0.8%; linalool oxide; nootkatone (1)
Active principles	Not known
Other chemical components	Not known
Products in which used	Non-alcoholic beverages
Level of use	No information available
Preparation	Tincture, essential oil
Main toxicological data	d-Limonene: ADI not specified (2)
	Linalool and linalyl acetate: ADI 0-0.5 mg/kg bw/d
Data needed	No data required
Specific observations	None
Classification and limits	**Fruit: category 1**
	Rind essential oil, tincture: category 1
National/int. evaluation	None
Main references	(1) J. Essent. Oil Res. 2:33 (1990)
	(2) JECFA 41st meeting (1993)
	(3) JECFA 23rd session (1979)
Data bases used	Chemical Abstracts 1956-95

Citrus japonica Thum.

CE No.	2032
Steinmetz No.	-
FEMA No.	-
Order	Rutales
Family	Rutaceae
Name	E Ronnel kumquat, round kumquat
	F Kumquat
	D Kumquat
	I Kumquat
	SP -
Synonyms	Fortunella japonica Swingle
Parts used	Fruit, rind
Important constituents	Rind essential oil: terpenic hydrocarbons [limonene 92-95%, β-myrcene 1.7-1.9%, α-pinene]; terpenic alcohols and esters [geraniol, nerol, linalool, α-terpineol, geranyl acetate, terpinyl acetate, neryl acetate]; carbonyle compounds [citral, decanal, citronellal, carvone]; aliphatic alcohols [hexanol, octenol-1] (1)
Active principles	Not known
Other chemical components	Not known
Products in which used	Beverages, candy
Level of use	No data available
Preparation	Rind essential oil
Main toxicological data	Limonene: ADI not specified (2)
Data needed	No data required
Specific observations	Fruit is used as a foodstuff
Classification and limits	**Fruit: category 1** **Rind essential oil: category 1**
National/int. evaluation	None
Main references	(1) CIVO-TNO – Volatile compounds in Food Ed. VI – p. 80-81 (2) JECFA 41st meeting 1993
Data bases used	Chemical Abstracts: 1956 – 1995 *Keywords:* Fortunella species, Kumquat, Kumquat peel oil

Citrus limon (L.) Burm.

CE No.	139 a
Steinmetz No.	314
FEMA No.	Lemon extract: 2623; Lemon oil: 2625; Lemon oil terpeneless: 2626; Lemon petitgrain oil: 2853
Order	Rutales
Family	Rutaceae
Name	E Lemon tree F Citronnier D Sauerzitronenbaum, Limonenbaum I Limone SP Limon
Synonyms	C. limonum Risso
Parts used	Fruit, pulp, rind, leaf and twigs
Important constituents	Rind expressed (or cold pressed) essential oil: (yield: 0,3-0,4%/fruit): limonene (54-76%), γ-terpinene (5-12%), α-and β-pinene, terpinolene, sabinene, myrcene, p cymene, camphene, bisabolene, caryophyllene; citral (2-3%), octanal, nonanal, decanal, dodecanal, citronellal; carvone; α-terpineol, linalool, octanol, geraniol, nerol, citronellol; geranyl acetate, neryl acetate, octyl acetate (1,2) + furocoumarins [psoralen, 5-methoxypsoralen (bergapten) 4-87 mg/kg, 8-methoxypsoralen (xanthotoxin), 5,8-dimethoxypsoralen (isopimpinellin), imperatorin, oxypeucedanin 26-728 mg/kg, phellopterin and 5-geranyloxypsoralen (bergamottin)] (3)
Active principles	Furocoumarins
Other chemical components	Not known
Products in which used	Alcoholic and non-alcoholic beverages, ices, candy, baked goods, gelatin and desserts, meat products
Level of use	Fruit juice: 400-1200 ppm Rind expressed oil: non-alcoholic beverages 160 ppm; alcoholic beverages 380 ppm; ices 395 ppm; candy 1156 ppm; baked goods 640 ppm; gelatin, desserts 534 ppm; meat products 500 ppm; others 4400 ppm Rind expressed terpeneless oil: non-alcoholic beverages 65 ppm; alcoholic beverages 44 ppm; ices 89 ppm; candy 136 ppm; baked goods 122 ppm; gelatin, desserts 147 ppm; meat products 50 ppm; others 2200 ppm Rind essential oil: non-alcoholic beverages 174 ppm; alcoholic beverages 249 ppm; ices 377 ppm; candy 562 ppm; baked goods 642

	ppm; gelatin, desserts 600 ppm; others 3000 ppm Rind essential terpeneless oil: non-alcoholic beverages 25 ppm; alcoholic beverages 26 ppm; ices 50 ppm; candy 125 ppm; baked goods 125 ppm; gelatin, desserts 50 ppm; others 2000 ppm Rind tincture: non-alcoholic beverages 575 ppm; alcoholic beverages 1030 ppm; ices 1375ppm; candy 4125 ppm; baked goods 3625 ppm; gelatin, desserts 300 ppm; others 1150 ppm (IOFI 1997)
Preparation	Juice, tincture, infusion, essential oil, "terpeneless essential oil", soluble oil
Main toxicological data	Low phototoxicity of lemon oil (4) d-Limonene: ADI not specified (5) Citral: ADI: 0,5 mg/kg bw/d (6) Furocoumarins: active principle (under evaluation)
Data needed	No data required
Specific observations	Fruit is used as a foodstuff
Classification and limits	**Juice, rind: category 1** **Rind essential oil: category 4 (with limits on furocoumarins)** **Rind infusion: category 1**
National/int. evaluation	Lemon extract: CFR 182.20 Lemon peel extract: CFR 182.20 Lemon oil: CFR 182.20 Lemon petitgrain oil: CFR 182.20
Main references	(1) J. Essent. Oil Res. 5 p. 21 (1993) (2) Perf.&Flav. 19 (3) p. 64 (1994) (3) Furocoumarins in Plant Food: TemaNord, 1996, 600, p.48 (4) Food Cosmet. Toxicol. 12, p.725 (1974) (5) JECFA 41st meeting 1993 (6) JECFA 23rd session, 1979
Data bases used	Chemical Abstracts: 1956-96 *Keywords:* Lemon, citrus limon

Citrus limon (L.) Burm.

CE No.	139 b
Steinmetz No.	314
FEMA No.	Leaf lemon oil (Petitgrain lemon oil): 2853
Order	Rutales
Family	Rutaceae
Name	E Lemon tree F Citronnier D Sauerzitronenbaum, Limonenbaum I Limone SP Limon
Synonyms	C. limonum Risso; C. medica var. limon L.; C. medica L. subsp. limon L.; C. medica var. limonum (Risso) Wight et Arn.
Parts used	Fruit, pulp, rind, 139b leaf and twig
Important constituents	Lemon leaf and twig essential oil: terpenic hydrocarbons [limonene 25.84-34.55%, α-pinene + α-thujene 0.83-2.15%, β-pinene 9.79-26.86%, sabinene 2.24-3.96% ,γ-terpinene 0.92-1.24%, β-ocimene 2.05-2.57%, myrcene 1.15-1.64%, b-caryophyllene 0.57-1.44%, δ-3-carene 0.57-0.98%]; terpenic alcohols and esters [linalool 1.24-1.46%, citronellol 0.27-0.39%, geraniol 0.99-1.63%, nerol 1.66-2.13%, α-terpineol 0.63-1.00%, geranyl acetate 2.30-3.31%, neryl acetate 4.12-8.18%]; carbonyle compounds [geranial 9.40-15.19% , neral 7.60-12.10%, citronellal 1.06-1.90%, nonanal 0.21-0.35%, methylheptenone 0.57-0.80%]; + methyl N-methyl anthranilate 0.14-0.19% (1)
Active principles	Not known
Other chemical components	Not known
Products in which used	Baked goods, frozen dairy, soft candy, gelatin, puddings, non-alcoholic beverages, alcoholic beverages (2)
Level of use	Leaf and twig essential oil: non-alcoholic beverages 1.8 ppm, alcoholic beverages 3.3 ppm, ices 5.3 ppm, others 25 ppm (5)
Preparation	Leaf essential oil
Main toxicological data	d-Limonene: ADI not specified (2) Citral (Geranial/Neral) ADI: 0-0.5 mg/kg/bw (3) Citronellal; linalool; geraniol; geranyl acetate: ADI: 0-0.5 mg/kg/bw (3) Geranyl acetate + citronellyl acetate: not carcinogenic in F344 rats and B6C3F1 mice (4)

Data needed	No data required
Specific observations	No relevant data found
Classification and limits	**Lemon leaf and twig essential oil: category 2**
National/int. evaluation	Lemon petitgrain oil: FDA 182.20
Main references	(1) Perf.&Flav., 18(5), p. 51, (1993) (2) JECFA 41st meeting 1993 (3) JECFA 23rd Session 1979 (4) NTP Technical Report No. 252, October 1987 (5) IOFI, 1997
Data bases used	Chemical Abstracts: 1956 -1996 *Keywords:* Lemon leaf oil, Petitgrain lemon oil

Citrus medica L. var. medica

CE No.	2035
Steinmetz No.	-
FEMA No.	-
Order	Rutales
Family	Rutaceae
Name	E Citron tree F Cédratier D Zitronatzitrone, Zitronatzitronenbaum I Cedro SP Cidro
Synonyms	Citrus medica (L.) var. macrocarpa Risso; Citrus medica (L.) var. vulgaris Risso; Citrus medica L. var. cedrata Risso
Parts used	Fruit, rind
Important constituents	Rind essential oil: terpenic hydrocarbons [limonene 60%, γ-terpinene 24%, α- and β-pinene 2%, myrcene 1.5%, α-thuyene, bergamotene, bisabolene], citral 3%; terpenic alcohols and esters: α-terpineol; geraniol; neryl and geranyl acetate (1)
Active principles	Not known
Other chemical components	Not known
Products in which used	Beverages, candy, crystallised fruit
Level of use	No information available
Preparation	Essential oil, infusion
Main toxicological data	d-Limonene: ADI not specified (2) Citral: ADI 0-0.5 mg/kg bw/d (3)
Data needed	Level of use
Specific observations	None
Classification and limits	**Fruit: category 1** **Rind essential oil: category 1**
National/int. evaluation	None
Main references	(1) Industrie delle Bevanda 17:7 (1988) (2) JECFA 41st meeting (1993) (3) JECFA 23rd meeting (1979)
Data bases used	Chemical Abstracts 1956-95 *Keywords:* Citrus medica oils

Citrus reticulata Blanco

CE No.	142
Steinmetz No.	317
FEMA No.	Mandarin oil: 2657, Petitgrain mandarin oil: 2854
Order	Rutales
Family	Rutaceae
Name	E Common mandarine tree, Suntara orange F Mandarinier, Mandarine D Mandarine, Mandarinenbaum I Mandarino SP Mandarina, mandarina petitgrain, aceite esencial
Synonyms	Citrus nobilis Andr. non Lour.
Parts used	Fruit, pulp, juice, rind, leaf and twig
Important constituents	Rind expressed (or cold pressed) essential oil: limonene 65-75%; γ-terpinene 16-22%; α- and β-pinene; myrcene; linalool; α-terpineol; α-sisensal 0.1-0.5%; methyl-N-methyl anthanilate 0.3-0.6% (1,2) Mandarin leaf oil (Japanese origin): limonene 1.2%; sabinene 23.6%; α-pinene 1.3%; β-pinene 1.4%; myrcene 2.2%; β-phellandrene 1.4%; (E)-β-ocimene 2.1%; p-cymene 0.6%; γ-terpinene 0.8%; linalool 59.2%; α-terpineol 1.2%; thymol and methyl thymol 0.3%; citral; geranyl acetate (3)
Active principles	Not known
Other chemical components	Not known
Products in which used	Non-alcoholic beverages, ice-cream, ices, candy, baked goods, gelatin and puddings, chewing gum, alcoholic beverages
Level of use	Mandarine oil expressed: baked goods 745.3 ppm; frozen dairy 251 ppm; soft candy 1250 ppm; gelatin, puddings 400 ppm; non-alcoholic beverages 36.45 ppm; alcoholic beverages 36.45 ppm; hard candy 1964 ppm; chewing gum 42.92 ppm (4) Petitgrain mandarin oil: baked goods 22.99 ppm; frozen dairy 8.12 ppm; soft candy 18.59 ppm; gelatin, puddings 15.02 ppm; non-alcoholic beverages 4.13 ppm; alcoholic beverages 9.27 ppm; hard candy 0.25 ppm (4)
Preparation	Essential oil, terpeneless essential oil, tincture, distillate, fluid extract
Main toxicological data	d-Limonene: ADI not specified (5) Linalool: ADI 0.5 mg/kg bw/d (6) Methyl-N-methyl anthranilate: ADI 0.2 mg/kg bw/d (6) Citral: ADI 0-0.5 mg/kg bw/d (6)

Data needed	No data required
Specific observations	None
Classification and limits	**Fruit: category 1** **Rind essential oil: category 1** **Leaf essential oil: category 2**
National/int. evaluation	CFR chap. 182.20, 580.20 FACC 1976 Appendix 1 (flesh) FACC 1976 Appendix 2 (leaves, rind, flowers)
Main references	(1) CIVO-TNO Volatile compounds in Food Ed. VI:73 (2) Lawrence, Essential oils (1979-91) (3) Perfumer&Flavorist 18:48 (1993) (4) Fenaroli Vol 1, p.181 (1995) (5) JECFA 41st meeting (1993) (6) JECFA 23rd session (1979)
Data bases used	Chemical abstracts 1956-96 *Keywords:* Mandarin oil, petitgrain mandarine oil

Citrus reticulata Blanco var. deliciosa H.H.Hu

CE No.	2039
Steinmetz No.	-
FEMA No.	Tangerine oil: 3041
Order	Rutales
Family	Rutaceae
Name	E Tangerine tree, mediterranean mandarin F Tangerinier, mandarinier, mandarin d'Italie D Tangerinenbaum, Italienischer Mandarinenbaum I Tangerino, Mandarino d'Italia SP
Synonyms	Citrus tangerina (Hort.ex Tan.); C. deliciosa Tenore; C. tangerina hort. ex Tanaka; Citrus nobilis Auct. non Lour
Parts used	Fruit, juice, rind, leaf and small twigs
Important constituents	Rind expressed (or cold pressed) essential oil: limonene 88-95%, α-terpinene 1.5-3.8%, α-pinene, β-myrcene, p-cymene, octanal, decanal, citral, α-sinensal, perillaldehyde; linalool, α-terpineol (1) Tangerine leaf oil (Dancy cultivar grown in Florida): α-pinene and thujene 1.1-2.7%, β-pinene 1.4-2.6%, myrcene 0.3-0.6%, α-terpinene 0.1-0.3%, limonene 0.8-1.9%, β-phellandrene 0.2-0.5%, ocimene [isomeric form not characterised] 2.8-8.2%, γ-terpinene 4.3-10%, methyl thymol 1.1-16%, α-terpineol and neral 0.2-0.6%, geraniol 0.1-0.2% (2)
Active principles	Not known
Other chemical components	Not known
Products in which used	Baked goods, frozen dairy, candy, gelatin, puddings, non-alcoholoc beverages, alcoholic beverages, chewing gum
Level of use	Baked goods 433.2 ppm; frozen dairy 319.2 ppm; soft candy 663.5 ppm; gelatin, puddings 190.2 ppm; non-alcoholic beverages 124.4 ppm; hard candy 499.0 ppm; chewing gum 723.3 ppm (3)
Preparation	Essential oil
Main toxicological data	d-Limonene: ADI not specified (4) Citral: ADI 0-0.5 mg/kg bw/d (group ADI) (5) Linolool: ADI 0.5 mg/kg bw/d (5) Thymol: metabolic study (6)
Data needed	No data required
Specific observations	None

Classification and limits	**Fruit: category 1** **Rind expressed oil: category 1** **Leaf oil: category 2**
National/int. evaluation	CFR chap 182.20, 580.20
Main references	(1) Lawrence: Essential oils 1976-7-9-80-1-7 (2) Perfumer&Flavorist 18:48 (1993) (3) Fenaroli vol 1, p 182, (1995) (4) JECFA 41st meeting (1993) (5) JECFA 23rd session (1979) (6) J Toxicol. Sci. 4:341 (1979), Pharmacol Toxicol 61:98 (1987)
Data bases used	Chemical Abstracts 1956-96 *Keywords:* Tangerine oil

Citrus reticulata Blanco var. unshiu (Marco.) H.H.Hu

CE No.	2031
Steinmetz No.	-
FEMA No.	-
Order	Rutales
Family	Rutaceae
Name	E Satsuma mandarine tree, Unshiu orange
	F Mandarinier Satsuma
	D Satsumamandarinenbaum, Satsumabaum
	I Mandarino giapponese
	SP Satsuma
Synonyms	Citrus unshiu Marc.; Citrus unshiu Makino; Citrus nobilis Lour. var. unshiu Swingle
Parts used	Fruit, juice, rind, leaf
Important constituents	Satsuma mandarin rind essential oil (or Mikan): D-limonene (86-90%), γ-terpinene, myrcene, p-cymene, linalool, α-terpineol. Satsuma mandarin leaf oil: γ-terpinene, p-cymene, camphene, limonene, β-caryophyllene, α- and β-elemene, β-bisabolene, linalool (1,2)
Active principles	Not known
Other chemical components	Not known
Products in which used	Beverages, candy
Level of use	No information available
Preparation	Juice, rind expressed oil, leaf oil
Main toxicological data	*d*-limonene: ADI not specified (3)
Data needed	Use levels of essential oils
Specific observations	None
Classification and limits	**Fruit: category 1**
	Rind essential oil: category 5
	Leaf essential oil: category 5
National/int. evaluation	None
Main references	(1) Perfumer&Flavorist 18:43 (1993)
	(2) CIVO-TNO-Volatile compounds in Food Ed.VI, pp 80-81
	(3) JECFA 41st meeting (1993)
Data bases used	Chemical Abstracts 1956-93
	Keywords: Mikan oil, citrus unshiu oil, satsuma mandarin oil

Citrus sinensis (L.) Pers.

CE No.	143
Steinmetz No.	318
FEMA No.	Orange cold pressed oil: 2825; orange distilled oil: 2821; orange oil terpeneless: 2822; orange peel extract: 2824; orange peel oil, terpeneless: 2826
Order	Rutales
Family	Rutaceae
Name	E Sweet orange tree F Oranger douce D Apfelsinenbaum, Orangenbaum I Arancio dolce SP
Synonyms	Citrus sinensis (L.) Osbeck; Citrus aurantium var. dulcis; Citrus aurantium var. sinensis
Parts used	Fruit, pulp, rind
Important constituents	Rind expressed (or cold pressed) essential oil: limonene 93.5-96.5%; myrcene 1.5-2%; α-pinene; γ-terpinene; linalool 0.4-1%; decanal <1.5%; octanal; citral <0.5%; α-sinensal (1,2,3) Rind infusion tincture: components of essential oil + bitter compounds
Active principles	Not known
Other chemical components	Not known
Products in which used	Alcoholic and non-alcoholic beverages, ice-cream, ices, candy, baked goods, gelatin and puddings, chewing gum, condiments, cereals, meat products, syrups, jams, jellies, gravies, sweet sauce, processed vegetables
Level of use	Orange distilled oil: baked goods 890.5 ppm; frozen dairy 156.6 ppm; meat products 24 ppm; condiment, relish 329.3 ppm; soft candy 1653 ppm; jam, jelly 125 ppm; sweet sauce 7500 ppm; gelatin, puddings 890 ppm; non-alcoholic beverages 106.9 ppm; alcoholic beverages 45.24 ppm; gravies 200 ppm; hard candy 2460 ppm; chewing gum 1989 ppm (4) Orange terpeneless oil: baked goods 40.61 ppm; frozen dairy 57.21 ppm; meat products 2 ppm; condiment, relish 2 ppm; soft candy 42.59 ppm; jam, jelly 40 ppm; gelatin, puddings 52.03 ppm; non-alcoholic beverages 58.76 ppm; alcoholic beverages 42.02 ppm; hard candy 122.9 ppm; chewing gum 387.5 ppm (4) Sweet orange peel extract: frozen dairy 400 ppm; soft candy 202.5

ppm; non-alcoholic beverages 140.6 ppm; alcoholic beverages 233.6 ppm; hard candy 399.5 ppm (4)

Sweet orange peel oil: baked goods 853.9 ppm; frozen dairy 370.7 ppm; meat products 23.0 ppm; condiment, relish 44.92 ppm; soft candy 717 ppm; gelatin, puddings 360.5 ppm; gravies 20 ppm; non-alcoholic beverages 166.7 ppm; alcoholic beverages 183.1 ppm; hard candy 1930 ppm; chewing gum 3817 ppm (4)

Sweet orange peel terpeneless oil: baked goods 75.43 ppm; frozen dairy 48.22 ppm; meat products 41 ppm; processed vegetables 2.3 ppm; condiment, relish 2 ppm; soft candy 25.9 ppm; confection, frosting 75 ppm; gelatin, puddings 48.53 ppm; non-alcoholic beverages 12.44 ppm; alcoholic beverages 49.21 ppm; hard candy 356.3 ppm; chewing gum 1483 ppm (4)

Preparation	Essential oils, terpeneless essential oil, soluble oil, infusion, tincture
Main toxicological data	d-Limonene: ADI not specified (5) Linalool; citral: ADI 0.5 mg/kg bw/d
Data needed	No data required
Specific observations	None
Classification and limits	**Rind: category 1** **Rind expressed oil: category 1** **Rind distilled oil: category 1**
National/int. evaluation	CFR chap. 182.20, 580.20
Main references	(1) CIVO-TNO Volatile compounds in Food (2) J Essent. oil Res. 6:101 (1994) (3) Perfumer&Flavorist 21(4):25 (1996) (4) Fenaroli Vol 1 p. 25 (1995) (5) JECFA 41st meeting (1993) (6) JECFA 23rd meeting (1979)
Data bases used	Chemical abstracts 1956-93 *Keywords:* orange oil, sweet orange oil

Citrus x paradisi Macfad.

CE No.	140
Steinmetz No.	315
FEMA No.	Grape fruit oil: 2530; Naringin extract: 2769
Order	Rutales
Family	Rutaceae
Name	E Common grapefruit tree F Grapefruitier, pamplemoussier D Grapefruit, Pampelmusenbaum I Pompelmo SP
Synonyms	C. paradisi Macf.; C. grandis (L.) Osbeck var. racemosa (Roem.) B.C.Stone; C.decumana (L.)
Parts used	Fruit, pulp, rind, leaf
Important constituents	Rind expressed (or cold pressed) essential oil (yield 0.06-0.08%): limonene 83-95%; myrcene; octanal; decanal; dodecanal; citral; geranyl acetate; neryl actate; nootkatone (1) Leaf essential oil: sabinene 42.0-59.0%; limonene 1.6-4.38%; α-pinene 1.6-2.98%; β-pinene 3.1-4.5%; myrcene 2.8-4.02%; linalool 5.9-24%; citronellol 1.4-3.83%; terpinen-4-ol 0.8-17.02%; citronellal 3.1-12%; geranial 0.6-1.7%; geranyl acetate 0.09-0.53% (2)
Active principles	Not known
Other chemical components	Not known
Products in which used	Baked goods, milk products, frozen dairy, soft candy, gelatin, puddings, non-alcoholic beverages, alcoholic beverages, hard candy, chewing gum (3)
Level of use	Grapefruit oil, expressed: Baked goods 860.7 ppm; milk products 9o ppm; frozen dairy 145.4 ppm; soft candy 1083 ppm; gelatin, puddings 883.3 ppm; non-alcoholic beverages 276.4 ppm; alcoholic beverages 132.8 ppm; hard candy 1977 ppm; chewing gum 1608 ppm (3) Naringin extract: baked goods 97.38 ppm; frozen dairy 54.23 ppm; soft candy 90.34 ppm; gelatin, puddings 78.16 ppm; non-alcoholic beverages 38.42 ppm; alcoholic beverages 175 ppm (3)
Preparation	Juice, essential oil, soluble oil, naringin extract
Main toxicological data	d-limonene: ADI not specified (4) Citral: ADI 0-0.5 mg/kg bw/d (group ADI) (5)

Data needed	Composition of naringin extract
	Use and use levels of leaf essential oil
Specific observations	None
Classification and limits	**Fruit: category 1**
	Rind and rind essential oil: category 1
	Leaf essential oil: category 5
	Naringin extract: category 5
National/int. evaluation	CFR chap. 182.20, 580.20
	FACC 1976 Appendix 2 (rind)
Main references	(1) Lawrence, Essential oils (1981-91)
	(2) Perfumer&Flavorist 18(5):56 (1993)
	(3) Fenaroli Vol 1, p.141 (1995)
	(4) JECFA 41st meeting (1993)
	(5) JECFA 23rd session (1979)
Data bases used	Chemical Abstracts 1956-93
	Keywords: Grapefruit oil, grapefruit leaf oil

Cocos nucifera L.

CE No.	147
Steinmetz No.	327
FEMA No.	-
Order	Principes
Family	Palmae
Name	E Coconut palm
	F Cocotier
	D Kokospalme
	I Albero del cocco
	SP Cocotero
Synonyms	Cocos mamillaris Blanco
Parts used	Fruit
Important constituents	Oil: acids: lauric acid (dodecanoic acid) 50%, myristic acid (tetradecanoic acid) 20%, palmitic acid (hexadecanoic acid) 10%, caprylic acid (octanoic acid) 9%, capric acid (deca-noic acid) 8%, stearic acid (octadecanoid acid) 3% (1, 2)
Active principles	Not known
Other chemical components	Not known
Products in which used	Commonly used in bakery products
Level of use	Annual use for flavouring purposes in Europe more than 1000 kg/year (IOFI) Use level of coconut oil in food 1 g/kg (IOFI 1994)
Preparation	Copra (the meat of the seeds) is removed and dried. It forms the shredded coconut used in cooking and pastry-making. Oil is obtained by extraction from the seeds (1)
Main toxicological data	Coconut oil caused toxic symptoms in young rats when incorporated in their diets at a level of 15% (3). Coconut milk in the Thai diet is considered as an important inhibitor of nonheme food iron absorption (4). Coconut products added to bakery products may produce acute allergic reactions (5)
Data needed	No data required
Specific observations	Cocos is used as a foodstuff
Classification and limits	Fruit: category 1
National/int. evaluation	None

Main references
(1) Tyler, V.E. et al. (1976)
(2) Duke, J.A. (1986)
(3) Liener, J.E. (1980)
(4) Am. J. Clin. Nutr. 28, 12:1348 (1975)
(5) Ann. Allergy 51, 4:472-81. (1983)

Data bases used
Medline (1966-92)
Toxline (1969-92)
Biosis (1973-92)
Chem. Abstr. (1969-92)
FSTA (1969-92)
Keywords: Latin and English names

Cola acuminata (P.Beauv.) Schott. & Endl.

CE No.	149
Steinmetz No.	-
FEMA No.	Cola nut extract: 2607
Order	Malvales
Family	Sterculiaceae
Name	E Kola
	F Cola
	D Kola
	I Cola
	SP Cola nuez, extracto
Synonyms	Cola pseudo-acuminata Engl., Sterculia acuminata P. Beauv.

Parts used	Seeds
Important constituents	Xanthines alkaloids [caffeine 2.4-2.6%, theobromine < 0.1%, theophylline]; flavonoids [d-catechol ~3%, colatin [= kolatin = l-epicatechol], colanin = kolanin = catechin-caffeine; anthocyanin pigment: kola red (= phlobaphen, oxidation product of catechols), organic acids: tannic acid, quinic acid, betaine, vitamins (1,2,3,4,5)
Active principles	Not known
Other chemical components	Caffeine
Products in which used	Baked goods, frozen dairy, soft candy, gelatin, puddings, non-alcoholic beverages, alcoholic beverages, hard candy
Level of use	Kola nut extract: baked goods 377,8 ppm, frozen dairy 445,8 ppm, soft candy 370 ppm, gelatin, puddings 370 ppm, non-alcoholic beverages 104,5 ppm, alcoholic beverages 133,8 ppm, hard candy 117,8 ppm (6)
Preparation	Tincture (20% in 60% ethanol), fluid, soft and dried extracts (also tannin-free).
Main toxicological data	Caffeine: review on metabolism, on toxicity, on carcinogenicity, on mutagenicity, on reproduction and prenatal toxicity, on consumption and on other biological effects in (7,8)
Data needed	No data required
Specific observations	Limits on caffeine should comply with national regulations
Classification and limits	**Kola and kola nut extract: category 4 (with limits on caffeine)**
National/int. evaluation	Kola nut, extract: CFR 182.20, 582.20; FEMA No.2607. In 1978, an FDA advisory panel concluded that caffeine, as it is added to cola

soft drink, should be subject to a more restrictive regulatory approach. Removal of Caffeine from the GRAS list " was urged ". (9)

Caffeine: IARC: inadequate evidence for the carcinogenicity in humans and in experimental animals. Not classifiable as to its carcinogenicity to humans (Group 3) (8)

Limits on caffeine in foods and beverages according to national regulations: France: limits of caffeine in foods and beverages (150 ppm)

Main references

(1) Crit. Rev. Toxicol., 5, p.189, (1977)
(2) Hoppe: Drogenkunde (1975)
(3) Rehm: Kulturpflanzen (1976)
(4) Food Chem. Toxicol., 27, p.49, (1989)
(5) Ann. Pharm. Fr., 44, p.495, (1986)
(6) Fenaroli, 1, p.161, (1995)
(7) Food Chem. Toxicol., 26, p.645, (1988); 30, p.533, (1992)
(8) IARC Monographs, 51, p.291, (1991)
(9) Duke: Handbook in Medicinal Herbs, 7th Printing, p.134, (1989)

Data bases used

Chemical Abstracts – 1997
Keywords: Cola acuminata

Cola nitida (Vent.) Schott. & Endl.

CE No.	2041
Steinmetz No.	-
FEMA No.	Cola nut extract: 2607
Order	Malvales
Family	Sterculiaceae
Name	E Kola-nut, Bitter cola nut
	F Cola
	D Kola
	I Cola
	SP Cola neuz
Synonyms	Cola acuminata (P. Beauv.) Schott & Endl. var. lalifolia K. Schum. Cola vera K. Schum.
Parts used	Seeds
Important constituents	Xanthines alkaloids [caffeine 1.5-3.5%, theobromine ~ 1%, theophylline]; primary and secondary amines [dimethylamine, methyl, ethyl-amine, isopentylamine]; flavonoids [d-catechine ~3%, colatine (l-epicatechine), colanine (catechin-caffeine); pigment: kola red (= phlobaphene), organic acids: tannic acid, quinic acid (1,2,3,4,5)
Active principles	Not known
Other chemical components	Caffeine
Products in which used	Kola nut: mainly in non-alcoholic beverages in USA
Level of use	Kola nut extract: baked goods 377,8 ppm, frozen dairy 445,8 ppm, soft candy 370 ppm, gelatin, puddings 370 ppm, non-alcoholic beverages 104,5 ppm, alcoholic beverages 133,8 ppm, hard candy 117,8 ppm (6)
Preparation	Tincture (20% in 60% ethanol), fluid, soft and dried extracts (also tannin-free)
Main toxicological data	Caffeine: see CE No. 149
Data needed	No data required
Specific observations	Limits on caffeine should comply with national regulations
Classification and limits	**Kola and kola nut extract: category 4 (with limits on caffeine)**
National/int. evaluation	Kola nut, extract: CFR 182.20, 582.20; FEMA No.2607 In 1978, an FDA advisory panel concluded that caffeine, as it is added to cola soft drink, should be subject to a more restrictive regulatory approach. Removal of Caffeine from the GRAS list " was

urged " (7)

Caffeine: IARC: inadequate evidence for the carcinogenicity in humans and in experimental animals. Not classifiable as to its carcinogenicity to humans (Group 3) (8)

Limits on caffeine in foods and beverages according to national regulations: [France: limits of caffeine in foods and beverages (150 ppm)]

Main references

(1) Crit. Rev. Toxicol., 5, p.189, (1977)
(2) Hoppe: Drogenkunde (1975)
(3) Rehm: Kulturpflanzen (1976)
(4) Food Chem. Toxicol., 27, p.49, (1989)
(5) Ann. Pharm. Fr., 44, p.495, (1986)
(6) Fenaroli, 1, p.161, (1995)
(7) Duke: Handbook in Medicinal Herbs, 7th Printing, p.134, (1989)
(8) IARC Monographs, 51, p.291, (1991)

Data bases used

Chemical Abstracts (1965 – 1997)
Keywords: Cola nitida

Corylus avellana L.

CE No.	155
Steinmetz No.	354
FEMA No.	-
Order	Fagales
Family	Betulaceae
Name	E Hazelnut tree
	F Noisetier
	D Haselnuss
	I Avellano, Avolano nocciolo, Bacuccola
	SP Avellano
Synonyms	-
Parts used	Nuts, leaves
Important constituents	Nuts: oil, containing myristic, oleic, linoleic and palmitic acid (1) and tocopherols (3); phytosterols, melibiose, manninotriose, raffinose and stachyose (1), filbertone (principal flavour) (6), phytate and polyphenol (4) Leaves: taraxerol (triterpene), beta-sitosterol and 3,7,22-alfa-trihy-droxystigmasten (sterol), myricitrin (flavonoid) and leaf aldehyde (1)
Active principles	Not known
Other chemical components	Not known
Products in which used	Nuts are used (often roasted) in confectionery and bakery products, or consumed in their natural form (5) Hazelnut oil (Oleum Corylus avellana) is used in confectionery and cosmetics. Leaves are used in herbal teas (1)
Level of use	Annual use of nuts in Europe 6200 kg (IOFI) Leaves 10g/l in beverages, not used in food (IOFI 1994)
Preparation	Oleum Corylus avellana
Main toxicological data	No relevant data found
Data needed	Quantitative data on chemical components in leaves and, if necessary, 28-day study and study on mutagenicity
Specific observations	None
Classification and limits	**Nuts: category 1** **Leaves: category 5**

National/int. evaluation None

Main references
(1) List & Hörhammer (1967-80)
(2) J. Allergy Clin. Immunol. 83:3 (1989)
(3) Annales pharm. francaises 30, 495-502 (1972)
(4) Am. J. Clin. Nutr. 47, 270-274 (1988)
(5) J. Food Sci. 37, 313-316 (1972)
(6) Angew. Chem. Int. Ed. Engl. 28:8, 1022-1023 (1989)

Data bases used
Chemical Abstracts (1967-90)
FSTA (1969-90)
Biosis (1973-90)
Medline (1966-90)
Toxline (1965-91)
Keywords: Corylus avellana L., hazel, hazelnut

Curcuma longa L.

CE No.	163
Steinmetz No.	371
FEMA No.	Powder: 3085; Extract: 3086; Oleoresin: 3087
Order	Scitamineae
Family	Zingiberaceae
Name	**E** Long rooted curcuma, Indian saffron, Turmeric plant
	F Curcuma longue
	D Gelbwurzel, Kurkuma
	I Curcuma di levante, Curcuma di indiane
	SP Curcuma
Synonyms	Curcuma domestica Val., Curcuma domestic Loir, Amomum curcuma Jacq.

Parts used	Rhizomes [cured and polished]
Important constituents	Essential oil of rhizomes contain turmerone (ca 60%), ar-turmerone, α- and γ-atlantone and zingiberene (25%) with minor amounts of eucalyptol, α-phellandrene, d-sabinene, borneol and dehydroturmerone. Also present is a yellow colouring matter comprising curcumin (0.3-5.4%), monosedmethoxycurcumin and didesmethoxycurcumin. Oleoresin contains curcumin (1, 2)
Active principles	Not known
Other chemical components	Eucalyptol
Products in which used	Curry powder, meats, condiments, gelatin, puddings, flavourings in food (eg bouillons, soups, sauces, pre-cooked beans), alcoholic beverages. Also used in egg products (1,2,3)
Level of use	Average maximum 22% in seasonings and flavourings, and 0.88% in condiments and relishes (2). Powder used at 30-8834 ppm; extract used at 40-59 ppm; oleoresin used at 3-2262 ppm (1)
Preparation	Tincture (20% in 60% ethanol), powder, oleoresin, oil, infusion, extract
Main toxicological data	JECFA temporary ADI for curcumin is 0.1mg/kg (June 89) (4). Subchronic oral studies in pigs with oleoresin showed increased liver/thyroid weights, from 60 mg/kgbw (5). Acute toxicity studies showed no effect at 2.5 g/kg in rats, guinea pigs, monkeys (6,7). Turmeric and curcumin negative in micronucleus test in rats, mice (8,9,10,11) and in Ames test (12,13). Turmeric, curcumin and turmeric oleoresin antimutagenic and anticlastogenic activity

	reported (14,15,16,17,18). Yellow pigment in turmeric has an irritant effect on the gastric mucosa (19)
Data needed	No data required
Specific observations	Used as a foodstuff
Classification and limits	**Essential oil: category 3 (with limits on eucalyptol)** **Rhizomes: category 3 (with limits on eucalyptol)**
National/int. evaluation	UK FACC (1976) Appendix 2. GRAS I and GRAS II. Fema Nos: 3085 (powder), 3086 (extract), 3087 (oleoresin).
Main references	(1) Fenaroli (1995) (2) Leung (1996) (3) Usher (1974) (4) JECFA (1990), 35th Report, WHO TRS No. 789 (5) Fd. chem. Tox. 23 (11) p 967-973 (1985) (6) Ind. J. Exp. Biol. 18, p73-4 (1980) (7) J. Fd Sci Tech. 19, p187-190 (1982) (8) Mut. Res. 79, p125-132 (1980) (9) Current Science, 55 (19), p1005-6 (1987) (10) Bionature 6(1) 7-10 (1986) (11) Mutat. Res. 136, 85-8 (1984) (12) Mut Research, 105 393-6 (1982) (13) Bull. Environ. Contam. Toxicol., 40, 350-357. (1988) (14) Food Chem. Toxic 25(7) 545-7 (1987) (15) Proceedings of AACR, 28, 173. (1987) (16) J. Nutrition, Growth and Cancer, 4, 83-89. (1987) (17) Cancer Letters, 29, 197-202. (1985) (18) Molecular and Cell. Biochem., 77(3),3-10. (1987) (19) Weiss, R. F. (ed.) Herbal Medicine (1988), Arcanum
Data bases used	Biosis (1973-90) Chemical Abstracts (1967-90) FSTA (1969-90) Toxline, Toxlit, Toxnet (1981-90) Toxlit65 (1965-90) Embase (1974-90) *Keywords:* Curcuma longa, Turmeric

Cymbopogon citratus (DC.) Stapf

CE No.	38
Steinmetz No.	89
FEMA No.	Lemongrass oil: 2624
Order	Graminales
Family	Gramineae
Name	E Lemongrass, West Indian
	F Lemongrass (de l'Amérique Centrale)
	D Lemongras, Zitronengras
	I Citronella (dell'America Centrale)
	SP Lemongras indias occidentales, extractos
Synonyms	Andropogon nardus var. ceriferus

Parts used	Herb, leaves (1)
Important constituents	The main components of the essential oil are neral (31%) and geranial (56-61%) (i.e. total citral content of 65-86%) and myrcene (12-20%; analgesic). Minor components are linalool (0.9-1.4%), methylheptenone (0.5-2.8%), geranyl acetate (1.1-1.2%) and traces (<1%) of eucalyptol, citronellal, α-thujone, α-terpineol, borneol, citronellol, nerol, geraniol and other compounds (2). In the leaf wax two triterpenoids cymbopogone and cymbopogonol (3), in various fractions of a light petroleum extract β-sitosterol, the two long-chain alcohols hexacosanol and triacontanol and a steroidal saponine have been isolated (4).
Active principles	Thujone
Other chemical components	Eucalyptol
Products in which used	Lemongrass is used as a spice in Europe and Southern Asia (5). Citral is an important starting material in the manufacture of ionones and ionone substitutes used extensively in flavours. Lemongrass oil is used in non-alcoholic and alcoholic beverages, frozen dairy, soft candies, baked goods, gelatin and puddings, and chewing gum (1)
Level of use	Lemongrass oil in non-alcoholic beverages 9.0 ppm, alcoholic beverages 8.9 ppm, frozen dairy 14 ppm, soft candy 33 ppm, baked goods 36 ppm, gelatin and puddings 19 ppm, chewing gum 197 ppm (1)
Preparation	Steam-distillation of freshly cut or partially dried grass (1,6). Infusion made by pouring boiling water on fresh or dried leaves (which is called "abafado" in Portuguese) (5)

Main toxicological data	Acute toxicity: "abafado" tea (up to 200 times the traditional human dosage of 2 g of dried leaves in 150 ml water) prepared from leaves of Brazilian Cymbopogon citratus (containing 46-99% citral and 16-36% myrcene) in rats and mice exerted no adverse effects and did not support the alleged folk use (anxiolytic, hypnotic) (5). Subchronic toxicity: no evidence of toxicity after "abafado" tea (20 times the traditional human dosage) daily for two months in rats (6). Human: no signs of toxicity and no effect with twice the traditional "abafado" tea doses for two weeks (7). Mutagenicity of lemongrass oil: negative in Bacillus subtilis (rec-/+ assay) (8). Tests on phototoxicity, irritation and sensitisation negative (9) β-Myrcene: Not cytotoxic and mutagenic in V79 cells and in mammalian cells in vitro. However, has antimutagenic properties (10) Citral: ADI 0-0.5 mg/kg bw/d (11)
Data needed	No data required
Specific observations	Lemongrass is used as a foodstuff
Classification and limits	**Herb: category 3 (with limits on eucalyptol and thujone)**
National/int. evaluation	Lemongrass oil: GRAS status (FEMA 1965); approved for food use by the FDA (GRAS)
Main references	(1) Fenaroli (1995) (2) Sci. Pharm. 51: 58-63 (1983) (3) Tetrahedon Letters 35: 3099-3102 (1975) (4) Planta Med. 28 (2): 186-189 (1975) (5) J. Ethnopharmacol. 17: 37-64 (1986) (6) J. Ethnopharmacol. 17: 65-74 (1986) (7) J. Ethnopharmacol. 17: 75-83 (1986) (8) J. Chem. Toxic. 20: 527-530 (1982) (9) Food Cosmet. Toxicol. 14: 455-457 (1976) (10) Environ. Molecular Mutagenesis 18: 28-34 (1991) (11) JECFA 23rd session (1979)
Data bases used	Toxline (1965-93) Chemical abstracts (1967-93) *Keywords:* Cymbopogon citratus, Lemongrass

Cymbopogon flexuosus (Nees ex Steud.) W. Wats.

CE No.	2045
Steinmetz No.	-
FEMA No.	-
Order	Graminales
Family	Gramineae
Name	E Lemongrass, East Indian
	F Lemongrass
	D Lemongras, Zitronengras
	I Lemongrass
	SP
Synonyms	Andropogon nardus var. flexuosus

Parts used	Herb, leaves (1)
Important constituents	The main components of the essential oil are neral (26-34%) and geranial (49-62%) (i.e. total citral 65-86-95%), besides small amounts of myrcene (0.2-0.6%), linalool (0.6-1.3%), limonene (0.1-2.4%), methylheptenone (0.5-1.4%), geranyl acetate (2%) and geraniol (3.8%) and trace (<1%) of α-terpineol, citronellol, nerol, citronellal and caryophyllene (2)
	Among the many distinct varieties and geographical species (over 50) the essential oil of some chemotypes of C. flexuosus (such as RRL-4, 22, 31, 57 and 59) has been found to be rich in sesquiterpenes (ca. 60%, mainly isointermedeol) and devoid of citral (3,4,5). Borneol (17% and 2%) and (+)-α-bisabolol (37% and 40%) were found as further major components in the chemotypes RRL-22 and 31, resp. (4). Trace amounts of methyleugenol (1-2%) have been reported in the essential oil of three of these chemotypes (RRL-4, 22 and 31) (4,5). An unusually high content of methyleugenol (20%) has been measured in the essential oil of the two other chemotypes (RRL-57 and 59) which also contained high amounts of geraniol (40% of oil) (6)
	The chemical and physical characteristics of Cymbopogon flexuosus (East Indian species) is very close to Cymbopogon citratus (West Indian species). The principal difference is the lower myrcene content in the East Indian species (7)
Active principles	Methyleugenol
Other chemical components	Not known

Products in which used	Used for flavouring purposes in non-alcoholic beverages, alcoholic beverages, frozen dairy, soft candy, baked goods, gelatin and puddings, and chewing gum (1)
Level of use	Lemongrass oil: non-alcoholic beverages 9.0 ppm, alcoholic beverages 8.9 ppm, frozen dairy 14 ppm, soft candy 33 ppm, baked goods 36 ppm, gelatin and puddings 19 ppm, chewing gum 197 ppm (1)
Preparation	Steam-distillation of freshly cut or partially dried grass (11, 12) Infusion made by pouring boiling water on fresh or dried leaves (which is called "abafado" in Portuguese) (8)
Main toxicological data	East Indian lemongrass oil: tests on phototoxicity, irritation and sensitisation negative (9) Citral: ADI 0-0.5 mg/kg bw/d (group ADI) (10)
Data needed	No data required
Specific observations	Lemongrass is used as a foodstuff. More than trace amounts of an active principle (methyleugenol) only present in one particular chemotype, therefore lemongrass can be classified in category 3
Classification and limits	**Herb, leaves: category 3 (with limits on methyleugenol)**
National/int. evaluation	Lemongrass oil: GRAS status (FEMA 1965); approved for food use by the FDA (GRAS)
Main references	(1) Fenaroli (1995) (2) Perfumer & Flavorist 19: 29-45 (1994) (3) J. Chromatogr. 262: 364-366 (1983) (4) J. Ess. Oil Res. 1: 107-110 (1989) (5) Phytochemistry 18: 671-672 (1979) (6) Indian J. Pharm. 38: 63 (1976) (7) Perfumer&Flavorist 3: 38-39 (1978) (8) J. Ethnopharmacol. 17: 37-64 (1986) (9) Food Cosmet. Toxicol. 14: 455-457 (1976) (10) JECFA 23rd sesstion (1979)
Data bases used	Toxline (1965-95) Chemical abstracts (1967-95) *Keywords:* Cympopogon flexuosus, Lemongrass

Cymbopogon martinii (Roxb.) W.Wats. var. martinii

CE No.	40
Steinmetz No.	91
FEMA No.	Palmarosa oil: 2831
Order	Graminales
Family	Gramineae
Name	E Palmarosa, East Indian Geranium
	F Palmarosa indien
	D Palmarosa
	I Palmarosa
	SP Palmarrosa, extractos
Synonyms	Andropogon martinii Roxb., Cymb. martinii W. Wats var motia, Cymbopogon martinii Stapf. var motia
Parts used	Herb (leaves, stems, flowering tops) (1)
Important constituents	Herb contains 0.3-1.0% essential oil. The main volatile components in the essential oil are geraniol (70-80%), geranyl acetate (5-20%), geranyl formate (5-15%), linalool (2-4%), limonene (0.2-2.2%), b-caryophyllene (1-3%), α-humulene (0.6%), estragole (traces) and further normal plant constituents in concentrations of < 1% (2,3,4,5)
Active principles	Estragole
Other chemical components	Not known
Products in which used	Palmarosa oil is used in beverages, candies, frozen dairy, puddings and baked goods (1, 6)
Level of use	Palmarosa oil used in non-alcoholic beverages 5.8 ppm, frozen dairy 10.0 ppm, soft candy 17 ppm, baked goods 20 ppm, alcoholic beverages 6.3 ppm, gelatin, puddings 9.2 ppm (1)
Preparation	Palmarosa oil is obtained by steam distillation of partially dried plants (1, 6)
Main toxicological data	Tests on irritation, sensitisation and phototoxicity of Palmarosa oil negative (6)
	Geraniol: Not mutagenic in Salmonella typhimurium TA 100 (7)
	Geranyl acetate: Not mutagenic in a rec-assay in Bacillus subtilis and to various strains of Salmonella typhimurium. Not carcinogenic in mice and rats in a two-year oral study at daily doses of 710 mg/kg bw/d and 1420 mg/kg bw/d, respectively. However, reduced survival and body weight gain, observed in high dose male rats and mice and high and low dose female mice, lowered the sensitivity of the

	study for detecting neoplastic responses. In male rats marginal increases of squamous cell papillomas in the skin and tubular cell adenomas of the kidney may have been related to geranyl acetate (7)
Data needed	No data required
Specific observations	None
Classification and limits	**Herb essential oil: category 4 (with limits on estragole)**
National/int. evaluation	GRAS status by FEMA 2831 and FDA. JECFA ADI: 0.5 mg/kg/ bw/d for geranyl acetate (8)
Main references	(1) Fenaroli (1995) (2) J. Agric Food Chem. 35 (1): 62-66 (1987) (3) Karrer et al., Konst. und Vorkommen der organ. Pflanzenstoffe, Ergänzungsband 1 (1977) (4) Phytochem. 26 (1): 183-185 (1987) (5) Perfumer&Flavorist 19: 29-45 (1994) (6) Food Cosmetic Toxicol. 12: 947 (1974). (7) NTP Technical Report 252 (1987) (8) JECFA (23rd meeting, 1979)
Data bases used	Chemical abstracts (1982-93) RTECS *Keywords:* Cymbopogon martinii, Palmarosa

Cymbopogon nardus (L.) W.Wats.

CE No.	39
Steinmetz No.	90
FEMA No.	Citronella oil: 2308
Order	Graminales
Family	Gramineae
Name	E Ceylon Citronella
	F Citronelle (de Ceylan)
	D Ceylon-Citronell
	I Citronella (di Ceylan)
	SP Citronela ceilan, extractos
Synonyms	Andropogon nardus (L.), Cymbopogon nardus (L.) W. Wats.
Parts used	Herb (1)
Important constituents	Ceylon citronella oil contains 55-65% total acetylizable alcohols (calculated as citronellol) and 7-15% total aldehyde (calculated as citronellal). The main constituents are geraniol (18-20%), citronellol (6.4-8.4%), citronellal (5-15%), geranyl acetate (2%), limonene (9-11%), methylisoeugenol (7.2-11.3%) (2,3,8,9). Others: camphene (7-8%), caryophyllene (0.9-3.2%), linalool (0.5-1.2%), citral (neral (0.4%) and geranial (0.7%)), methylheptenone (0.7%), methyleugenol (1.7%), l-borneol (5-7%), nerol (0.6-0.9%), eugenol (traces), farnesol (traces) (9)
Active principles	Methyleugenol
Other chemical components	Not known
Products in which used	Citronella oil is used to flavour beverages, candies, frozen dairy, puddings and baked goods (1, 4)
Level of use	Citronella oil used in food: non-alcoholic beverages 26.23 ppm, frozen dairy 42.58 ppm, soft candy 46 ppm, baked goods 48 ppm, gelatin, puddings 39 ppm, breakfast cereals 29 ppm, alcoholic beverages 40 ppm, hard candy 13 ppm (1)
Preparation	Ceylon Citronella oil is obtained by steam-distillation of the partially dried herb known as the lenabatu variety (1)
Main toxicological data	Tests on irritation and sensitisation of Ceylon citronella oil negative (4)
	Gosselin et al. (6) reported that citronellol, a major constituent of citronella oil, produced paralysis, coma and death in doses of about 1-4 ml/kg given by stomach tube in rabbits. Mant (7) reported a human exposure case in which vomiting, shock, cyanosis and con-

vulsions preceded death in a child who consumed an unknown quantity of a commercial preparation consisting largely of oil of citronella, however it is questionable whether the fatal outcome of this case was solely due to oil of citronella toxicity. Other cases of citronella oil intoxication of children ended without major symptoms (5)
Methyleugenol: active principle

Data needed	No data required
Specific observations	Herb is used as a foodstuff
Classification and limits	**Herb: category 3 (with limits on methyleugenol)** **Essential oil: category 3 (with limits on methyleugenol)**
National/int. evaluation	Citronella oil: GRAS status by FEMA 2308 and FDA, JECFA ADI: 0.5 mg for citral, citronellal and citronellol (8)
Main references	(1) Fenaroli (1995) (2) Karrer W., Konst. und Vorkommen der organ. Pflanzenstoffe, (1976) (3) Karrer et al., Konst. und Vorkommen der org. Pflanzenstoffe, Ergänzungsband 1, (1977) (4) Food Cosmetic Toxicol., 11: 1067 (1973) (5) Clinical Toxicol. 29 (2): 257-262 (1991) (6) Clinical Toxicol. of Commercial Prod. II, 231 (1984) (7) Med Sci Law, 1/2: 170-171 (1961) (8) JECFA (23rd meeting, 1979) (9) Perfumer & Flavorist, 19: 29-45 (1994)
Data bases used	Chemical abstracts (1982-93) RTECS *Keywords:* Cymbopogon nardus, Citronella

Cymbopogon winterianus Jowitt

CE No.	2046A
Steinmetz No.	-
FEMA No.	Citronella oil: 2038
Order	Graminales
Family	Gramineae
Name	E Java Citronella
	F Citronelle (de Java)
	D Java-Citronellgras
	I Citronella (di Java)
	SP
Synonyms	-
Parts used	The freshly cut or partially dried herb (1)
Important constituents	Java citronella oil contains not less than 35% alcohols (calculated as citronellol) and not less than 35% aldehydes (calculated as citronellal). The main constituents are citronellal (32-45%), geraniol (21-24%), citronellol (11-15%), geranyl acetate (3-8%), limonene (1.3-3.9%), elemol and other sesquiterpene alcohols (2-5%) and among others small amounts of linalool (0.8%), eugenol (0.6%), caryophyllene (0.4-2.1%), citral (neral (0.6%) and geranial (0.8%)), nerol (0.3%), camphene (0.04%) (2)
Active principles	Not known
Other chemical components	Not known
Products in which used	Citronella oil is used to flavour beverages, candies, frozen dairy, puddings, and baked goods (1, 3)
Level of use	Herb essential oil: non-alcoholic beverages 5 ppm, alcoholic beverages 7 ppm, ices 5 ppm, candies 20 ppm, baked goods 15 ppm, desserts 5 ppm, meat products 5 ppm, soups 3 ppm, snacks 5 ppm (IOFI 1998)
Preparation	Java Citronella oil is obtained by steam or water distillation of the freshly cut or partially dried herb known as the mahapengira variety (1)
Main toxicological data	Tests on irritation and sensitisation of Java citronella oil negative (3) Citronellol, a major constituent of citronella oil, produced paralysis, coma and death in doses of about 1-4 ml/kg given by stomach tube in rabbits (4). The reason for the death in a reported case of citronella oil intoxication of a child with fatal outcome is unclear (5).

	Other cases of citronella oil intoxication of children ended without major symptoms (6)
Data needed	No data required
Specific observations	Herb is used as a foodstuff
Classification and limits	**Herb: category 1** **Oil: category 1**
National/int. evaluation	Citronella oil: GRAS status by FEMA 2308 and FDA; JECFA ADI: 0.5 mg for citral, citronellal and citronellol (7)
Main references	(1) Fenaroli (1995) (2) Perfumer & Flavorist, 19, 29-45, (1994) (3) Food Cosmet. Toxicol., 11, 1067ff, (1973) (4) Clin. Toxicol. of Commer. Prod. II, 231, (1984) (5) Med. Sci. Law, 1/2, 170-171, (1961) (6) Clin. Toxicol., 29 (2), 257-262, (1991) (7) JECFA (23rd meeting, 1979)
Data bases used	Chemical abstracts (1982-96) Medline (1966-98) Embase (1980-98) Biological Abstracts (1989-98) CC Life (2/97-2/98) *Keywords:* Citronella, Cymbopogon winterianus

Eucalyptus globulus Labill.

CE No.	185
Steinmetz No.	442
FEMA No.	Eucalyptus oil: 2466
Order	Myrtiflorae
Family	Myrtaceae
Name	E Eucalyptus tree
	F Eucalyptus officinal, eucalyptus globuleux
	D Eukalyptus, Fieberbaum, Blaugummibaum
	I Eucalipto
	SP Eucalipto, aceite esencial
Synonyms	Eucalyptus glauca DC

Parts used Leaves, flowers

Important constituents Leaves essential oil from ssp. globulus: terpenic hydrocarbons [a-pinene 3.1-9.7%, b-pinene 0.1-0.3%, myrcene trace-0.2%, p-cymene 0.7-1.9%, limonene 0.70-4.23%, aromadendrene 0.3-1.2%, allo-aromadendrene trace-0.3%, α-cubebene 0.25%, β-gurjunene 0.84%, α-gurjunene 0.44%, δ-cadinene 0.42%]; oxygenated terpenic compounds [1,8-cineole 62.4-82.2%, linalool 0.1-0.5%, α-fenchyl alcohol 0.1-1.5%, *trans*-pinocarveol 1.4-4.3%, isopulegol trace-2.2%, pinocarvone 0.3-0.4%, borneol 0.5-1.8%, terpinen-4-ol 0.3-0.6%, p-cymen-8-ol 0.1-1.1%, α-terpineol 1.5-2.9%, myrtenal 0.1-0.5%, *cis*-carveol 0.2-0.5%, *trans*-carveol 0.5-1.2%, piperitone trace-0.2%, nerolidol 0.2-1%, ledol 0.1-0.4%, spathulenol 1.4-6.9%, globulol 0.3-1.6%, guaiol 0.1-1%, eudesmol (3 isomers) 0.2-0.4%]; ester: a-terpinyl acetate 1.2-2.8% (1,2, 3)

Leaves essential oil from ssp. maldenii F. Muell.: terpenic hydrocarbons [α-pinene 0.5-5.83%, β-pinene trace-0.23%, myrcene trace-0.2%, p-cymene trace-0.8%, limonene 0.4-1%, aromadendrene 0.3-1.5%, allo-aromadendrene 0.1-0.4%, α-terpinene 0.03-1.15%, β-phellandrene 5.5%, γ-terpinene 0.04-1.45%]; oxygenated terpenic compounds [1,8-cineole 68.9-80.2%, linalool 0.1-0.4%, α-fenchyl alcohol 0.1-1.5%, *trans*-pinocarveol 0.6-4.3%, isopulegol trace-0.9%, pinocarvone 0.2-0.8%, borneol 0.6-2.6%, terpinen-4-ol 0.2-0.8%, p-cymen-8-ol 0.1-0.9%, α-terpineol 1.8-6.9%, myrtenal trace-0.9%, *cis*-carveol 0.2-0.5%, *trans*-carveol 0.7-1.5%, piperitone trace-0.2%, α-terpinyl acetate trace-0.4%, nerolidol 0.4-0.9%, ledol 0.2-0.6%, spathulenol 1.8-5.8%, globulol 0.6-3.6%, guaiol 0.2-0.9%, eudesmol (3 isomers) 0.1-0.3%] (1,2, 3)

	Flowers (=buds) oil: terpenic hydrocarbons [α-thujene 11.95%, β-pinene 0.3%, sabinene 0.54%, myrcene 0.07%, limonene 3.1%, p-cymene 8.04%, α-cubebene 0.34%, aromadendrene 16.57%, allo-aromendrene 1.88%, α-copaene 0.08%, β-bourbonene 1.57%, viridiflorene 0.36%]; oxygenated compounds [1,8-cineole 36.95%, pinocarvone 0.1%, cryptone 0.11%] + ester: isoamyl isovalerate 0.05% (4)
Active principles	Not known
Other chemical components	Eucalyptol
Products in which used	Infusion (2%), tincture (20% in 80% ethanol), fluid extract, soft water extract, essential oil
Level of use	Eucalyptus oil: baked goods 10.47 ppm, frozen dairy 5.39 ppm, meat products 18.02 ppm, condiment, relish 4 ppm, soft candy 9.40 ppm, gelatin, puddings 7 ppm, non-alcoholic beverages 2.17 ppm, alcoholic beverages 2.07 ppm, hard candy 1958 ppm, chewing gum 5.31 ppm (5)
Preparation	Leaves: Infusion (2%), tincture (20% in 80% ethanol), fluid extract, soft water extract, essential oil
Main toxicological data	Eucalyptol (1,8-cineole): toxicity studies: (1)- Hybrid B6C3F1 mice: (a) 28-day gavage (doses 0 – 150 – 300 – 600 and 1200 mg/kg bw): no dose-related lesions detected. (b) 28-day encapsulated form in the feed (concentrations 3750 – 7500 – 15000 and 30000 ppm): minimal hypertrophy of centrilobular hepatocytes, not considered to be significant pathology (6) Fischer 344 rat: (c) 28-day gavage (doses 0 -150 – 300 – 600 -1200 mg/kg): NOAEL 300 mg/kg. (d) 28-day encapsulated feed study (concentrations 0 – 7500 – 15000 – 30000 ppm): treatment-related renal lesions found only in males (7). Induced accumulation of protein droplets containing $\alpha_2\mu$-globulin in proximal tubular cells in male Wistar rats given 1000 mg 1,8-cineole/kg body wt/day for 28 days (8). Main urinary metabolites in brushtail possum (*Trichosurus vulpecula*) = 9-hydroxycineole and cineol-9-oic acid (9,10). In rats treated by gavage main metabolites = 2-hydroxycineole, 3-hydroxycineole (neutral metabolites) and 1,8-dihydroxy-10-carboxy-p-menthane (acidic metabolite) (11) [formation of this acidic metabolite requires the hydrolysis of the cyclic ether linkage]. In rabbits given eucalyptol by gavage same urinary neutral hydroxylated metabolites: 2- and 3-hydroxycineole (12). Mutagenicity tests negative with *Bacillus subtilis* rec-assay (13), with CHO cells (no increase in SCE's) (14)
Data needed	No data required
Specific observations	None
Classification and limits	**Eucalyptus leaves, flowers and preparations: category 4 (with limits on eucalyptol)**

National/int. evaluation	Eucalyptus oil: CFR:172.510
Main references	(1) J. Essent. Oil Res., *8*, p.19, (1996)
	(2) Perf. Flav., 22(1), p.49, (1997)
	(3) J. Essent. Oil Res., *9*, p.159, (1997)
	(4) J. Essent. Oil Res., *7*, p.147, (1995)
	(5) Fenaroli, *1*, p.121, (1995)
	(6) 28-day gavage and encapsulated feed study on 1,8-cineole in B6C3F1 hybrid mice NTP Experiment No. 5014-03 (encapsulated) and 5014-07 (gavage) – NCTR Experiment No. 389, 440, National Center for Toxicological Research, Jefferson (Arkansas 72079), April 1987
	(7) 28-day gavage and encapsulated feed study on 1,8-cineole in Fischer 344 rats NTP Experiment No. 5014-02 (encapsulated) and 5014-06 (gavage) – NCTR Experiment No. 380, 439, National Center for Toxicological Research, Jefferson (Arkansas 72079), April 1987
	(8) Toxicology Lett., *80*, p.147, (1995)
	(9) Aust. J. Chem., *32*, 2093, (1979)
	(10) Xenobiotica, *10*, 17, (1980)
	(11) Bull. Environ. Contam. Toxicol., *37*, 759, (1986)
	(12) J. Agric. Food Chem., *37*, 222, (1989)
	(13) J. Osaka City Med. Cent., *34*, 267, (1986)
	(14) Mutat. Res., *226*, 103, (1989)
Data bases used	Chemical Abstracts, Toxline (1965- 1997) *Keywords:* Eucalyptus globulus

Evernia prunastri (L.) Ach.

CE No.	194
Steinmetz No.	467
FEMA No.	Absolute: 2975
Order	Lecanorales
Family	Parmeliaceae
Name	E Oak Moss
	F Mousse de chêne
	D Eichenmoos
	I Muschio di quercia
	SP Musgo de encina, extractos
Synonyms	-
Parts used	Sucker/lichen
Important constituents	Extract of lichen contains α- and β-thujone, camphor, depsides (evernic acid, usinic acid, evernin, atranorin, chloroatranorin, lecanorin), orcinol, atranol, sparassol, divarine and other monaryl derivatives of depsides (1,2). Monoterpenes e.g. α and β-pinene (2). Sesquiterpenes e.g. α-humulene and β-caryophylene, fatty acid esters, hydrocarbons (2)
Active principles	Thujone
Other chemical components	Camphor
Products in which used	Absolute used in the formulation of aromas (1,3). Also used in flavourings of fruit, honey, spice (4)
Level of use	Non-alcoholic beverages: 3.5 ppm; alcoholic beverages 4 ppm; ice-creams: 1 ppm; confectionery: 1.6 ppm; baked goods: 1.4 ppm; gelatin desserts: 1.2 ppm; soups: 0.5 ppm(1). In UK used in condiments from 0.01 – 0.5 ppm
Preparation	Extract, concrete, absolute, resinoid
Main toxicological data	Extracts of Oak Moss cause contact sensitivity. Atranorin is one of the components responsible for this effect (5,6). Camphor is extremely toxic when ingested, causing CNS excitement, delirium, muscular excitability and epileptiform convulsions. Adult human lethal dose is in the range 50-500mg/kg
Data needed	Quantitative data on chemical components and, if necessary, an oral 28-day study and mutagenicity studies on the relevant extract
Specific observations	None

Classification and limits	Sucker: category 5 (with limits on camphor and thujone)
National/int. evaluation	UK FACC (1976) Appendix 2 Oak moss absolute FEMA No. 2795; FDA 172.510 Thujone-free. Thujone use restricted as in Annex II of Council Directive 88/388/EEC
Main references	(1) Fenaroli, 1995 (2) International Congress of Essential Oils (Pap) 7th, Vol. 7:384-7 (1979) (3) Arctander 1960 (4) MAFF (1995) (5) Contact Dermatitis 9,3: 227 (1983) (6) Contact Dermatitis 7: 168 (1981)
Data bases used	Biosis (1995) Chemical Abstracts (1995) FSTA (1969-) Toxline, Toxlit, Toxnet (1995) Toxlit(1995) Embase (1995) *Keywords:* Evernia prunastri, Parmelia prunastri, Labaria prunastri, Oakmoss

Ferula assa-foetida L.

CE No.	196
Steinmetz No.	473
FEMA No.	Asafetida fluid extract: 2106; Asafetida gum: 2107; Asafetida oil: 2108
Order	Umbelliflorae
Family	Umbelliferae
Name	E Asafetida F Ase fétide D Stinkasant, Teufelsdreck I Assa fetida SP Asafetida
Synonyms	Ferula assafoetida L.
Parts used	Gum resin = Gommo-oleoresin
Important constituents	Composition of gommo-oleoresin: resin: 40-60%; gum: 25-48%; volatile compounds: 7-12% (1) Volatile compounds from gommo-oleoresin (essential oil): mainly disulfides: 2-butyl-1-propenyl disulfide: 36-84% (E and Z; E/Z = 7/3); two diastereoisomers of 1-(1 methyl thio propyl)-1-propenyl disulfide: 9-31%; two diastereoisomers of 2-butyl-(3-methylthio-2-propenyl) disulfide: 0-32%; low amounts of 2-butyl-ethyl disulfide; dimethyl trisulfide; 2-butyl methyl disulfide; di-2-butyl disulfide; di-2-butyl trisulfide; di-2-dibutyl tetrasulfide; di-1-butyl trisulfide; 2-butyl ethenyl disulfide; n-butyl ethyl disulfide; 2-butyl-1-butenyl disulfide (E and Z) (ref 2,3,4); terpenic hydrocarbons (α-and β-pinene; phellandrene; δ-3-carene; asaresene A; camphene; p-cymene) (2,3,4,5,6,7) Composition of resin: asaresinotanol free or combined with ferulic acid; farnesiferols A, B or C and allied compounds (3); umbelliferone; valeric acid; traces of vanilline (2) Composition of gum: mainly polysaccharides; glucose; galactose; L-arabinose; rhamnose; glucuronic acid (1,3)
Active principles	Not known
Other chemical components	Not known
Products in which used	Non-alcoholic beverages; ices and icings; candy; baked goods; meats; condiments; sauce ("Worcestershire sauce"); snack foods; gravies
Level of use	(8) Asafetida gum: baked goods: 11.06 ppm; meat products: 2.25 ppm; soft candy: 10 ppm

Asafetida fluid extract: baked goods: 29.36 ppm; fats, oils: 20 ppm; frozen dairy: 20 ppm; meat products: 43.79 ppm; condiments, relish: 58.13 ppm; soft candy: 16 ppm; snack foods: 20 ppm; non-alcoholic beverages: 15 ppm; gravies: 6 ppm

Asafetida oil: baked goods: 12.95 ppm; frozen dairy: 10 ppm; meat products: 49.68 ppm; condiment, relish: 3.02 ppm; soft candy: 1.41 ppm; gelatin, puddings: 15 ppm; snack foods 1 ppm; non-alcoholic beverages: 15 ppm; foods 0.5-5 ppm (IOFI, 1994)

Preparation	Essential oil; fluid extract; tincture (20% in 70% ethanol)
Main toxicological data	No relevant data found
Data needed	Composition of fluid extract 28-day oral study of essential oil
Specific observations	None
Classification and limits	**Gommo-oleoresin: category 5** **Essential oil: category 5** **Fluid extract: category 5**
National/int. evaluation	Asafetida fluid extract: FDA 182.20 Asafetida oil: FDA 182.20
Main references	(1) Leung, p.37, (1980) (2) Perf.&Flav., 6(2), p.38, (1981); 9(2), p.43, (1985) (3) Indian Food Packer, p.29, (Jan/Feb 1979); p.65, (Sep/Oct 1982) (4) Phytochemistry, 23, p.899, (1984) (5) Pakistan J. Sci. Ind. Res., 23, p.68, (1980) (6) Planta Med., 53, p.300, (1987) (7) Acta Chem. Scand., B30, p.137, (1976) (8) Fenaroli, 1, p.39, (1995)
Data bases used	Chemical Abstracts – 1991; Pascal – 1991 *Keywords:* Ferula

Ferula gummosa Boiss.

CE No.	197
Steinmetz No.	474
FEMA No.	-
Order	Umbelliflorae
Family	Umbelliferae
Name	E Galbanum tree
	F Férule à galbanum, galbanum
	D Galbanum, Mutterharzbaum
	I Galbano
	SP Galbano
Synonyms	Ferula galbaniflua Boiss. & Buhse
Parts used	Gommo-oleoresin (Galbanum)
Important constituents	Gommo-oleoresin: resin 60%; gum 30-40%; essential oil 5-26%
	Gum: polysaccharides; galactose; arabinose; galacturonic acid; 4-methylglucuronic acid; umbelliferone (1); disulfides (sec. butyldisulfide; isopropenldisulfide) (2)
	Resin: resinic acids (1); sesquiterpenic compounds: α-terpenyl acetate; α-fenchyl acetate; guaiol; bulnesol; β-eudesmol; (Z) and (E) dihydrofarnesol; 10 epi-elemol; (E)guai-9-en-11-ol; (Z) dihydrofarnesol; α- and β-dihydroagarofuran; 10 *epi*-junenol; epi-ligulyloxide (3,4) and macrolides (2)
	Absolute: sesquiterpenic compounds: 10-*epi*-junenyl acetate; shyobunol; shyobunyl acetate; *epi*-shyobunol; *epi*-shyobunyl acetate and macrolides; 12-tridecanolide; 13-tetradecanolide; 14-pentadecanolide; 15-hexadecanolide (5,6)
	Essential oil: terpenic hydrocarbons (65-75%)[b-pinene 45-50%; δ-3-carene 10-20%; α-pinene 10-20%; limonen terpinolene; camphene; (E) and (Z) ocimeme; myrcene (0.5-1%)(1,4)]; terpenic alcohols and esters (<1%): linalool and linalyl acetate; α-fenchol and fenchyl acetate (2,6,7); sesquiterpenic hydrocarbons (8-10%) mainly cadinene (7); sesquiterpenic alcohols mainly guaiol; bulnesol; α-, β- and γ-eudesmol (8); azulene derivatives (1); thiol-esters (<0.1%)(1,8); pyrazine derivatives (<0.1%)(1,8); disulfides [(E) and (Z) propenyl disulfide; isobutyl disulfide (8)]; (E,Z) and (E,E)1,3,5-undecatriene (responsible for the odour of the oil) (7)
Active principles	Not known
Other chemical components	Not known

Products in which used	Alcoholic and non-alcoholic beverages; ices and icings; candy; baked goods; gelatin and puddings; condiments; meats; sauces
Level of use	Galbanum essential oil: baked goods 16 ppm; frozen dairy 10.47 ppm; meat products 22.40 ppm; condiment, relish 13.68 ppm; soft candy 14 ppm; gelatin puddings 15.50 ppm; snack foods 0.05 ppm; non-alcoholic beverages 7.88 ppm; alcoholic beverages 11 ppm; gravies 24 ppm; hard candy 0.07 ppm Galbanum resin: baked goods 30 ppm; frozen dairy 21 ppm; meat products 8 ppm; soft candy 33 ppm; gelatin puddings 33 ppm; non-alcoholic beverages 11 ppm (8)
Preparation	Essential oil; resin
Main toxicological data	Galbanum oil: slightly irritating. Tested at 4% in petroleum, no irritation after a 48 hr closed patch test on human subjects (9) Galbanum resin: no irritation when tested at 4% in petroleum after a 48 hr closed patch test on human subjects (9) Main components of galbanum oil: α- and β-pinene: in man α-pinene has caused skin irritation, and skin sensitisation, particularly when allowed to oxidise. The vapour induced eye, nose and throat irritation. α-pinene has been given orally in combination with several other terpenes to treat gallstones. α-pinene was of low acute oral and dermal toxicity in the rat and rabbit respectively. Lethal oral doses apparently produced local irritation, central nervous system depression and respiratory failure in rats, while repeated administration at low levels caused liver enzyme induction. Liver and kidney damage was found in sheep given several high oral doses. Oral administration to pregnant rats of a product containing α-pinene, plus several other terpenes was foetotoxic only at a maternally toxic dose level. α-pinene aparently had a weak tumour-promoting effect on mouse skin when repeatedly applied after a single dose of a known carcinogen, but it was not mutagenic in Ames bacterial tests (10)
Data needed	28-day oral oral study on α- and β-pinene. 28-day oral study and mutagenicity studies on α- and β-carene
Specific observations	None
Classification and limits	**Galbanum oil: category 5** **Galbanum resin: category 5**
National/int. evaluation	None
Main references	(1) Leung, p 175 (1980) (2) Augereau J.M., Sanofi-Elf-Bio-Recherches, Labège, p 21 (1987) (3) Perfumer&Flavorist, 7:21 (1983) (4) Helv.Chim.Acta 61:2874 (1978) (5) Perfumer&Flavorist 4:53 (1979) (6) Helv.Chim.Acta 61:2671 (1978) (7) Bull.Soc.Chim.(France)1:97 (1967) (8) Fenaroli p 131 (1995)

(9) Food Cosmet. Toxicol. 16:765 (1978)
(10) BIBRA Toxicity Profile: a-pinene (1992)

Data bases used   Chemical Abstracts 1967-95

Fortunella japonica (Thunb.) Swingle

CE No.	2032
Steinmetz No.	-
FEMA No.	-
Order	Rutales
Family	Rutaceae
Name	E Kumquat
	F Kumquat
	D Kumquat
	I Kumquat
	SP
Synonyms	Fortunella margarita; Citrus japonica Thun.
Parts used	Fruit, rind
Important constituents	Rind essential oil: limonene 92-95%, b-myrcene 1.7-1.9%, α-pinene; geraniol, nerol, linalool, α-terpineol, hexanol, octenol 1; citral decanal, ditronellal; carvone; geranyl acetate, terpinyl acetate, neryl acetate (1)
Active principles	Not known
Other chemical components	Not known
Products in which used	Beverages, candy
Level of use	No information available
Preparation	Rind essential oil
Main toxicological data	d-Limonene: ADI not specified (2)
Data needed	Level of use of essential oil
Specific observations	None
Classification and limits	**Fruit: category 1** **Rind essential oil: category 5**
National/int. evaluation	JECFA ADI: 0.5 mg/kg bw/d for citral (3)
Main references	(1) CIVO-TNO-Volatile compounds in Food Ed. VI pp 80-81 (2) JECFA 41st meeting (1993) (3) JECFA 23rd meeting (1979)
Data bases used	Chemical abstracts 1956-91 *Keywords:* Kumquat peel oil, Fortunella species

Galium odoratum (L.) Scop.

CE No.	77
Steinmetz No.	-
FEMA No.	-
Order	Gentianales
Family	Rubiaceae
Name	E Sweet woodruff
	F Asperule odorante, petit muguet
	D Waldmeister, Maikraut
	I Asperella odorata
	SP Hepatica estrellada
Synonyms	Asperula odorata L.
Parts used	Herb
Important constituents	Dried herb contains about 1% coumarin (min.-max. 0.7-1.7%; mean content in April/May 1.06% and in August 0.44-0.93%) (1). In fresh herb the iridoids asperulosid (0.28%), monotropein (0.05%; 0.25% in dried herb) (2), scandoside (0.022%; 0.05% in stem) and traces of deacetylasperuloside (0.0013%) and deacetylasperuloside acid (0.0025%) (3). Phenolic compounds: gallic acid, caffeic acid, p-coumaric acid, p-hydroxybenzoic acid, vanillin. Others: n-alkanes C_{19}-C_{31} (0.024% of dry wt) of which n-C_{27} (55%) and n-C_{29} (29%) are dominating (4,5) There are 224 volatile compounds reported in extract (pentane/dichloromethane 2+1) from dried herb (6)
Active principles	Coumarin
Other chemical components	Not known
Products in which used	Infusion of herb is used in beverages, alcoholic beverages, ices/frozen dairy, candies, and baked goods (IOFI 1997)
Level of use	Infusion of herb: non-alcoholic beverages 250 ppm, alcoholic beverages 900 ppm, ices 500 ppm, candies 500 ppm, baked goods 750 ppm (IOFI 1997)
Preparation	Infusion (IOFI 1997), dried herb, alcoholic extract
Main toxicological data	Headache and daze may occur after consumption of high amounts of 'Maiwein' prepared with fresh herb. It is recommended not to use more than 3 g of fresh herb to prepare 1 liter of such an aromatic wine (1) Coumarin: Coumarin is hepatotoxic in rats (NOEL 1000 mg/kg feed, liver damage at 2600 ppm). Toxic signs in dog after 100 mg/kg bw

orally (8d; vomiting, depression, impaired liver function). Reports on carcinogenic potential in rats not conclusive. Relevance for humans remains unclear, as there are known interspecies differences in metabolism. In contrast to rats and dogs, the main part of the absorbed coumarin is metabolised to 7-hydroxycoumarin in humans. Only 1-6% is metabolised into the potentially hepatotoxic o-hydroxyphenylic acid, which is a main metabolite in rats. In human, urinary route is the primary route of excretion, whereas in rats biliary extretion is of a high order. (7). Risk of long-term toxicity regarded as low. Reproduction: retardation of fetal development in mice after 0.25% coumarin in diet (8)

In humans no adverse reactions after 100 mg/d oral, no effect on bone marrow at 400-3500 mg/d (10). Mutagenicity: microsomal assay with S. typhimurium positive at 1 mg/plate, DNA damage at 20 mmol/l in mammal lymphocytes (9), Ames test negativ with TA1535, 1537, 98 and 100 (7). Reproduction: Fetotoxicity TDL_0 mouse oral 3600 mg/kg (6-17d preg) (9); Teratogenicity: no malformations in offspring of mice at 0.05-0.25% in feed (6-17d preg) (7)

A coumarin content of 5 mg/l in spiced wine (Maibowle) is considered as safe for human consumption (1)

Coumarin: active principle

Data needed	No data required
Specific observations	None
Classification and limits	Herb: category 4 (with limits on coumarin)
National/int. evaluation	SCF-opinion on coumarin: general limit for food and beverages recommended at the currently achievable limit of detection of 0.5 mg/kg (SCF, Dec 1996)
	The use for the preparation of essences is not allowed in Germany according to the "Aromenverordnung" (BGBl.I. p.1625, 1677). The use in 'Maiwein' (aromatic wine flavoured with woodruff) is limited to a coumarin content of 5 mg/l (according to German regulation: "Verordnung über Wein, Likörwein und weinhaltige Getränke" of 15th July 1971 (BGBl.I. p. 926), modified on 20th July 1977 (BGBl.I. p.1416))
	CE 1974 permitted presence of 5 ppm in food and 10 ppm in alcoholic beverages. In the USA, the bureau of Alcohol, Tobacco and Firearms, limited coumarin in alcoholic liquors to 5 ppm (7)
Main references	(1) Deutsche Apotheker Zeitung 17: 848-850 (1985)
	(2) Pharm. Acta Helv. 46: 121-128 (1971)
	(3) Krumholz B., Dissertation, Universität Frankfurt/Main (1988)
	(4) Haenseler R. et al., Hager's Handbuch der Pharmazeutischen Praxis. 5th Ed., Springer Verlag, Berlin (1990)
	(5) Phytochemistry 17: 1131-1133 (1978)
	(6) Z. Lebensm. Unters. Forsch 193: 317-320 (1991)
	(7) Fd. Cosmet. Toxicol 17: 277-289 (1979)
	(8) Planta medica 41: 221-231 (1981)

(9) RTECS, Registry of Toxic Effects of Chemical Substances. US Dept of Health and Human Services, Washington (1987)
(10) Teuscher E. et al., Biogene Gifte – Biologie, Chemie, Pharmakologie. 2nd Ed., G. Fischer, Stuttgart (1994)

Data bases used Medline (1966-97)
Embase (1980-97)
Toxline (1965-97)
Biological Abstracts (1989-97)
Keywords: Galium odoratum, Asperula odorata, sweet woodruff

Gardenia jasminoides Ellis

CE No.	210
Steinmetz No.	-
FEMA No.	-
Order	Gentianales
Family	Rubiaceae
Name	E Garden gardenia, Cape jasmin
	F Gardenia
	D Gardenia
	I Gardenia
	SP
Synonyms	G. florida L., G. radicans Thunb., G. grandiflora Lour., G. augusta Merr.

Parts used	Flowers, Fruits (1, 2)
Important constituents	Fruits: a number of iridoid glycosides: gardenoside, geniposide (genipin-1-glucoside), shanzhiside, genipin-1-B-D-gentiobioside, gardoside, scandoside methyl ester, geniposidic acid. Further the pigment crocetin (8,8'-diapo-ψ,ψ-carotenedioic acid), its glycoside crocin, and a glycosidic bitter substance, picrocrocinic acid. The lipoxygenase inhibitors 3,4-dicaffeoyl-5- (3-hydroxy-3-methylglutaroyl) quinic acid and 3-caffeoyl-4-sinapoylquinic acid. Leaves: a formyl substituted iridoid, cerbinal Flowers: steroidal compound gardenoic acid B. Main constituents of the essential oil obtained from the flower were benzyl acetate, hydroxycitronellal, and eugenol (3). Presence of sitosterol is also mentioned (1). An old handbook mentions the presence of styrolylacetate (responsible for the scent) and linalool (2). The major components of the absolute obtained from the flowers were linalool (12.0%), *cis*-3-hexenol (1.0%), palmitic acid (7.2%) and jasmin lactone (8.0%; mainly responsible for the sweet odour of the flower), hydrocarbons such as α-farnesene (18.0%), trans-ocimene (1.4%) and squalene (3.0%), aliphatic and other esters such as methyl benzoate (1.0%), ethyl 5-hydroxy-*cis*-7-decenoate (3.0%), *cis*-3-hexenyl tiglate (10.0%), hexyl tiglate (1.0%), and *cis*-3-hexenyl benzoate (5.0%) (4)
Active principles	Geniposide
Other chemical components	Not known

Products in which used	Use as a flavouring in beverages, ices, candies, baked goods and gelatin puddings (IOFI 1997). Used as a colour additive in various kinds of foods and beverages (5)
Level of use	Tincture of the flowers: beverages 118 ppm, gelatin desserts 270 ppm. Essential oil of the flowers: non-alc. beverages 10 ppm, alc. beverages 60 ppm, ices 60 ppm, candies 60 ppm, baked goods 60 ppm, gelatin desserts 60 ppm and other products 10 ppm (IOFI 1997)
Preparation	For flavouring purposes: essential oil, tincture (IOFI 1997); for traditional use: fluid extract, dried fruit, dry-fried fruit (1)
Main toxicological data	Hepatotoxic activity of α-naphthylisothiocyanate was significantly reduced by geniposide administered orally to rats (6). The extract of the fruit showed no hepatotoxic effects in rats (3). Another study reported hepatotoxicity of the crude extract of gardenia fruit after acute oral exposure in rats (7). It was shown that geniposide exerted hepatotoxic effects when administered orally at doses of ≥320 mg/kg bw, but not after i.p. application. Genipin, the aglycon, was hepatotoxic only after i.p. and not oral application with comparable effects after doses ≥80 mg/kg bw. These results indicate that genipin has a key role in the hepatotoxicity caused by geniposide (8) Crocin at 0.1 g/kg given i.v. increased bile secretion in rabbits. Crocin at 50 mg/kg given orally for 8 days had no effect on hepatic function of rats. A high dose of 100 mg/kg for 2 weeks induced acute hepatic damage and black pigmentation, which were both reversible. No effect with 10 mg/kg (3). Gardenoic acid B showed a significant effect on terminating early pregnancy in rats (3). Genotoxicity: negative Ames test with S. typhimurium TA98 and TA100, negative Rec assay with Bacillus subtilis (9)
Data needed	No data required
Specific observations	None
Classification and limits	**Flowers (tincture, oil): category 4 (with limits on geniposide)**
National/int. evaluation	None
Main references	(1) Bensky, D. et al., Chinese Herbal Medicine – Materia Medica, rev. Ed., Eastland Press, Seattle, (1993) (2) Burkill, I.H., Dictionary of the economic products of the Malay Peninsula. London, (1935) (3) Tang, W. et al., Chinese Drugs of Plant Origin. Springer, Berlin, (1992) (4) Agric. Biol. Chem. 42 (7): 1351-1356 (1978) (5) Carcinogenesis 7 (4): 595-599 (1986) (6) Active component from a Chinese composite prescription for the treatment of liver diseases. In: Chang, H.M. et al. Advances in Chinese Medicinal Materials Research. World Scientific, Singapore (1985) (7) Toxicology Lett. 44: 177-182 (1988)

(8) Fd. Chem. Toxic. 28 (7): 515-519 (1990)
(9) Mutat.Res. 97: 81-102 (1982)

Data bases used   Medline (1966-97)
Embase (1980-97)
Toxline (1965-97)
Biological Abstracts (1989-97)
Keywords: Gardenia jasminoides, Gardenia florida

Gentiana lutea L.

CE No.	214
Steinmetz No.	-
FEMA No.	-
Order	Gentianales
Family	Gentianaceae
Name	E (Yellow) Gentian, Bitter wort, Common gentian
	F Gentiane jaune, grande gentiane
	I Genziana, Genziana maggiore
	D Gelber Enzian
	SP Genciana raices, extracto
Synonyms	Asterias lutea Borkh., Swertia lutea Vest
Parts used	Roots (of two-year-old plant), herb (1, IOFI 1997)
Important constituents	The dried roots contain small amounts (2-3%) of secoiridoidglycosides: gentiopicrin (syn. gentiopicroside, was also called "gentiamarin") as main part of secoiridoid fraction, swertiamarin, and esters of sweroside and swertiamarin. Bitter taste mainly due to the acylglycoside amarogentin (0.2-0.5% of which 91% is found in the rind (2,3). Furthermore yellow xanthones [ca. 0.1%; gentisin (1,7-dihydroxy-3-methoxyxanthone, also known as gentiamarin) and isogentisin (1,3-dihydroxy-7-methoxyxanthone) as major components, gentisein (1,3,7-trihydroxyxanthone), methylgentisin and other 1,3,7-polyoxygenated derivatives, free and as glycosides i.e. gentisin-1-O-primveroside and gentioside-7-O-primveroside (4,5), also the precursor to gentisein 2,3',4,6-tetrahydroxy-benzophenone (0.003% in fresh roots) (5)]; phytosterols, phenolic acids, oligosaccharides (30-50%; sucrose, the trisaccharide gentianose (2.5-5%) (6), the disaccharide gentiobiose (5-8%; bitter), inulin and pectin-like polysaccharides (3-11%), small amounts of essential oil (0.002-0.007%)(2) with main components limonene (34.7%), linalool, carvacrol, cis-linalyloxide and α-terpineol totalling 50%. Fractions with most intense and typical gentian flavour are those containing aldehydes such as octen-2-al, nonen-2-al, decen-2-al, undecen-2-al, decadien-2,4-al and β-cyclocitral (7). No starch present (8)
	The leaves contain the secoiridoid gentiopicrin, the xanthones gentiosid, isogentisin and mangiferin, and the flavonoid heterosides isoorientin-4'-O-glucoside and isovitexin-4'-O-glucoside (3,9)
	Amarogentin is found in the capsules. Reported pyridin-alkaloides of the type of gentianine are artefacts caused by isolation procedures (6)
Active principles	Xanthones

Other chemical components	Not known
Products in which used	Used as an aperitive in gentian bitters, liqueurs and "angostura bitters" and in non-alcoholic beverages with bitter taste (8). Also used in ices, candies, baked goods, gelatin desserts and other products (IOFI 1997)
Level of use	Herb: tincture in non-alcoholic beverages 1000 ppm; essential oil in non-alcoholic beverages 10 ppm, alcoholic beverages 60 ppm, ices 60 ppm, candies 1030 ppm, baked goods 60 ppm, gelatin desserts 60 ppm and other products 1030 ppm. Roots: Tincture in non-alcoholic beverages 1200 ppm, alcoholic beverages 50,000 ppm, ices 1938 ppm, candies 122 ppm, baked goods 1417 ppm, gelatin desserts 125 ppm and other products 1298 ppm; concrete in non-alcoholic beverages 69 ppm, alcoholic beverages 69 ppm and candies 6 ppm; essential oil in non-alcoholic beverages 10 ppm, alcoholic beverages 60 ppm, ices 60 ppm, candies 1030 ppm, baked goods 60 ppm, gelatin desserts 60 ppm and other products 1030 ppm (IOFI 1997)
Preparation	Infusion (2%), fluid extract, soft and dried aqueous extracts, and tincture (20% in 60% ethanol), essential oil (1, IOFI 1997)
Main toxicological data	The drug may not be well tolerated by those with very high blood pressure or by expectant mothers (10). Headache may occur in sensitive persons (11) Genotoxicity: positive Rec assay with B. subtilis (water extract); positive Ames test with S. typhimurium TA100 (negative with TA98) with (water and methanol extract) (12,13) and/or without metabolic activation (water extract) (12). Mutagenic potential due to the xanthones gentisin and isogentisin, which are responsible for 76% of total mutagenic activity of methanol extract (14). Ames test with isolated xanthones positive with TA100 (+S9) for gentisein, gentisin and isogentisin. Isogentisin showed the highest mutagenic activity and was also positive with TA100 (15)
Data needed	No data required
Specific observations	None
Classification and limits	**Roots and preparations: category 4 (with limits on xanthones)** **Herb and preparations: category 4 (with limits on xanthones)**
National/int. evaluation	None
Main references	(1) Fenaroli (1995) (2) Planta Med. 28: 289-300 (1975) (3) Haenseler, R. et al. Hager's Handbuch der Pharmazeutischen Praxis. 5th Ed., Springer Verlag, Berlin (1990) (4) Karrer, W. Konstitution und Vorkommen der organischen Pflanzenstoffe. 2nd Ed., Birkhäuser, Basel (1976) (5) Tetrahedron 25: 1507-1511 (1969)

(6) Steinegger, E. et al. Pharmakognosie. 5th Ed., Springer Verlag, Berlin (1992)
(7) Z. Lebensm. Unters. Forsch. 182: 212-214 (1986)
(8) Bruneton, J. Pharmacognosy, Phytochemistry, Medicinal Plants. Lavoisier, Paris (1995)
(9) Helv. Chim. Acta 56:3050-3054 (1973)
(10) Tyler, V.E. The Honest Herbal – A Sensible Guide to the Use of Herbs and Related Remedies. George F. Stickley, Philadelphia (1982)
(11) Wichtl, M. Teedrogen. 2nd Ed., WVG, Stuttgart (1989)
(12) Mutat. Res. 97: 81-102 (1982)
(13) Genet. Toxicol. Diet 206: 63-72 (1986)
(14) Mutat. Res. 116: 103-117 (1983)
(15) Mutat. Res. 150: 141-146 (1985)

Data bases used

Medline (1966-97)
Embase (1980-97)
Toxline (1965-97)
Biological Abstracts (1989-97)
Keywords: Gentiana lutea, Yellow Gentian

Hierochloe odorata L.

CE No.	230
Steinmetz No.	562
FEMA No.	-
Order	Graminales
Family	Gramineae
Name	E Holy grass, Sweet grass
	F Herbe de la Ste-Vierge
	D Mariengras, Büffelgras
	I Avena odorata, (Erba della Sta. Vergine)
	SP -
Synonyms	-
Parts used	Herb (1), roots (2)
Important constituents	The essential oil contains coumarin (10-60%, i.e. ethanol extract (EtOH): roots 10%, aerial parts 25%; pentane-ether extract (P-E): roots 29%, aerial parts 62%) (2,3,4), 3-methylbutanal (=isovaler-aldehyde; 38% in roots (EtOH)), 3-methylbutanol (0.7-3.3% in roots (EtOH)), furfural (5% in whole plant (EtOH)) and aliphatic acid ethyl esters as main constituents (2). A characteristic minor constituent is massoia lactone (5,6-dihydro-6-pentyl-2H-pyran-2-one or dec-2-en-5-olide; 1.4% in whole plant (P-E)) (2,5). Yield of 2-3% dried ethanol extract from fresh roots and herb. Content of essential oil in roots 0.004-0.008% and in herb 0.0009-0.0057% (extracted with EtOH or P-E prior to distillation) (2)
Active principles	Coumarin
Other chemical components	Not known
Products in which used	Herb is used in non-alcoholic and alcoholic beverages, ices, candies, desserts, meat products and soups (IOFI 1998). Herb is traditionally used in the production of a special kind of vodka, i.e. Subrowka, and also for flavouring foods with natural coconut-like notes (massoia lactone) (5)
Level of use	Herb absolute: non-alcoholic beverages 0.3-3.0 ppm, alcoholic beverages 0.6 ppm, ices 0.01-20 ppm, candies 7-30 ppm, baked goods 0.01 ppm, desserts 0.01-20 ppm; herb as such: alcoholic beverages 800-5000 ppm; herb extract: non-alcoholic beverages 10 ppm, meat products 10 ppm, soups 100 ppm; herb distillate: alcoholic beverages 130-6800 ppm (IOFI 1998). (IOFI 1998).
	Coumarin content of up to 11.8 mg/l in vodka and up to 2.5 mg/l in liqueurs flavoured with Hierochloe odorata reported (6)

Preparation	Essential oil: steam-distillate of alcoholic extract of roots or aerial parts (2). Herb absolute and extract (IOFI 1998)
Main toxicological data	Massoilactone is a defence substance also isolated from ants. It is a powerful skin irritant, and produces systolic standstill in frog heart muscle (7)
Data needed	No data required
Specific observations	None
Classification and limits	**Herb, roots and preparations: category 4 (with limits on coumarin)**
National/int. evaluation	None
Main references	(1) CoE, Flavour. Subst. and Nat. Sources of Flavourings, Strasbourg (1981) (2) Flavour&Fragrance J., 6: 63-68 (1991) (3) Karrer W., Konst. und Vorkommen der org. Pflanzenstoffe (1976) (4) Karrer et al., Konst. und Vorkommen der org. Pflanzenstoffe, Ergänzungsband 1 (1977) (5) Flavour and Fragrance J., 5: 71-73 (1990) (6) Mitt. Rebe-Wein-Obstbau u. Früchteverw., 24 (5) (1974) (7) Aust. J. Chem. 21: 2819-2823 (1968)
Data bases used	Chemical abstracts (1967-93) Medline (1966-97) Embase (1980-97) Toxline (1965-97) Biological Abstracts (1989-97) *Keywords:* Hierochloe odorata, Holy grass

Hordeum vulgare L.

CE No.	232
Steinmetz No.	566
FEMA No.	-
Order	Commelinanae
Family	Gramineae (Poaceae)
Name	E Barley
	F Orge
	D Gerste
	I Orzo
	SP Cebada
Synonyms	H. sativum Pers., H. hexasticum L.
Parts used	Seeds (1)
Important constituents	Barley contains gramine, hordenine (alcaloid), N-methyl-tyramine, o- and p-coumaric acid, ferulic acid, starch, proteins (hordein proteins, prolamin), neutral-, glyco- and phospholipids; malt contains about 50-70 % maltol, 2-15 % dextrins and 8 % proteins (2,3,4,5,6)
Active principles	Not known
Other chemical components	Not known
Products in which used	Generally used as starchy foodstuffs and in brewing for the preparation of malt and beer, used in flavours (1)
Level of use	Not known
Preparation	Malt or malted barley: by artificially germinating dried barley grains; malt extract: by extracting malt with water (5)
Main toxicological data	No relevant data found
Data needed	No data required
Specific observations	Barley as a cereal is toxic to cocliaic patients
Classification and limits	**Seeds: category 1**
National/int. evaluation	Malt extract: GRAS (unpublished) (1)
Main references	(1) Fenaroli (1995)
	(2) Karrer W., Konst. und Vorkommen der organ. Pflanzenstoffe (1976)

(3) Karrer et al., Konst. und Vorkommen der organ. Pflanzenstoffe, Ergänzungsband 1, (1977)
(4) Phytochemistry 27 (7): 2037-2039 (1988)
(5) Tyler V.E., Pharmacognosy (1988)
(6) Canad. J. Plant Sci. 60: 1343 (1980)

Data bases used Toxline (1981-91)
FSTA (1970-90)
Chemical Abstracts (1987-91)
Biosis (1973-90)
Keywords: Barley, Hordeum vulgare L.

Hypericum perforatum L.

CE No.	234
Steinmetz No.	576
FEMA No.	-
Order	Guttiferales
Family	Guttiferae
Name	E St. John's wort, Klamath weed
	F Millepertuis, herbe de St-Jean
	D Johanniskraut
	I Iperico
	SP Hipericon
Synonyms	-
Parts used	Flower tops, herb, leaves and caulis (1, 2)
Important constituents	Whole herb contains vitamin C (0.13%), tannin, myrcene, α-pinene, alkaloids, ocimene (3, 4) and xanthone derivatives (5). Contains red pigment hypericin, whole herb (0.0095-0.466%)(3, 4), flowers (16.2%)(6) also, seed contains hypericin (4). (NB Level of Hypericin varies depending on the country of origin and state of the plant before extraction (ie whether plant is dried or wet etc.)) (7). Seeds contain hyperin (4). Flowers contain hyperoside, hyperforin, pseudohypericin, quercetin (6). Essential oil contains 2-methyloctane (16.4%), α-pinene (10.6%), dodecanol (5%), nonane (3.4%), 3-methyl nonane (3.2%) undecane (3.2%), isoundecane (3.1%), 6-methyl-5-hepten-2-one (2.1%) (8)
Active principles	Hypericin, xanthones
Other chemical components	Not known
Products in which used	Dried products used in formulation of liqueurs (2). Leaves and fruit used to make tea (9)
Level of use	No information available
Preparation	Infusion [3%], fluid extract and tincture [20% in 25% EtOH] (1)
Main toxicological data	8 Long-Evans rats fed dried St John's Wort in diet at 5% for 178 days, no tissue lesions, organs not specified in 4 rats at day 119, no effect on survival or liver mineral level at day 178 (10). Photosensitivity induced in calves with single dose 3g/kg of dried plant (hypericin 0.054 mg/g dried plant). NOEL of 1 g/kg was observed in this study (7). Plant extract inhibits monoamine oxidase (MAO) activity in vitro (11, 5) Ethanolic extract non-mutagenic in vitro and in vivo (12). Procyanidin fraction had a vasodilator effect

	on precontracted porcine coronary arteries (13). Hypericum tincture (0.25 ml/100g bw) produced a significant anti-inflammatory effect (48%) on formalin induced arthritis in albino rats (14).
Data needed	Use level of flavouring preparations, hypericin level in preparations. 28-day oral study including examination of inhibitory effect of extracts on MAO activity & clarification whether hypericin or xanthone derivatives are responsible
Specific observations	None
Classification and limits	Herb: category 5 (with limits on hypericin and xanthones)
National/int. evaluation	UK FACC (1976) Appendix 2 Hypericin use restricted as in Annex II of Council Directive 88/388/EEC
Main references	(1) Fenaroli (1995) (2) Fenaroli (1975) (3) MAFF (1995) (4) Duke (1985) (5) Pharmacopsych. 22(5): 194 (1989) (6) Holzl et al. (1987) Deut. Apoth Zeit 127, 23, p1227-1230 (1987) (7) J. Comp. Path 91, p135-141 (1981) (8) J. High Resol. Chrom & Chrom. commun. 5, (April),7 , p182-188 (1982) (9) Usher (1974) (10) Tox Lett. 10, p183-188 (1982) (11) Plan. Med. 46, p272-274 (1984) (12) Arzneim. Forsch. 40, 8, 851-5 (1990) (13) Arzneim. Forsch./Drug Res. 41 (i), No.. 5, 481-83.(1991) (14) British Homeopathic Journal, 77, 27-29. (1988)
Data bases used	Biosis (1973-91) Chemical Abstracts (1967-91) FSTA (1969-91) Toxline, Toxlit, Toxnet (1981-91) Toxlit65 (1965-91) Embase (1974-91) *Keywords:* Hypericum perforatum, St. Johnswort, John's wort

Illicium verum Hook.

CE No.	238
Steinmetz No.	583
FEMA No.	Star anise: 2095; Star anise oil: 2096
Order	Magnoliales
Family	Lauraceae
Name	E Star anise
	F Anis étoilé
	D Sternanis
	I Anice stellato
	SP Badiana
Synonyms	-
Parts used	Seed
Important constituents	Star anise oil: terpenic compounds [α-pinene 0.02-0.57%, camphene 0.03- 0.05%, β-pinene 0.03-0.05%, sabinene 0.03-0.14%, α-phellandrene 0.08-0.39%, p-cymene 0.09-0.29%, myrcene 0.12-0.41%, limonene 1.18-2.38%, γ-terpinene 0.01-0.30%, (Z)-β-ocimene 0.32-0.43%, (E)-β-ocimene 0.07-0.09%]; alkenyl benzenes [trans-anethole 71.19-87.89%, cis-anethole 0.20-0.42%, safrole trace-0.14%, estragol (=methyl chavicol) 0.34-5.04%]; oxygenated derivatives [linalool 0.42-0.73%, terpinen-4-ol 0.03-0.38%, α-terpineol 0.11-0.38%, nerolidol 0.12-0.80%, anisaldehyde 0.26-4.48%, foeniculin 0.50-0.89%?] esters [cinnamyl acetate trace-0.25%, methyl anisate trace-0.09%] (1,2)
Active principles	Estragole and safrole
Other chemical components	Not known
Products in which used	Various food products
Level of use	Star anise: baked goods 150 ppm, frozen dairy 20 ppm, meat products 1000 ppm, soft candy 90 ppm, gelatin, puddings 15 ppm, non-alcoholic beverages 15 ppm, alcoholic beverages 60 ppm (3)
	Star anise oil: baked goods 38.05 ppm, frozen dairy 4.46 ppm, meat products 0.32 ppm, soft candy 260.9 ppm, gelatin, puddings 3.08 ppm, non-alcoholic beverages 15.18 ppm, alcoholic beverages 167.4 ppm (3)
Preparation	Fluid extract, tincture (20% in 60 to 70% ethanol), essential oil (source of trans-anethole)
Main toxicological data	trans-anethole: 2 years chronic/carcinogenic study in rats: NOEL 105-125 mg/kg/day (4). Review on genotoxicity studies (5).

	Metabolic study in man (6). DJA: 0-2 mg/kg bw, JECFA 51st meeting, 1998 Safrole: active principle
Data needed	No data required
Specific observations	Chinese star anise should not be confused with Japanese star anise (Illicium lanceolatum A.C.Smith), which is highly poisonous
Classification and limits	**Star anise and star anise oil: category 3 (with limits on estragole and safrole)**
National/int. evaluation	Star anise: CFR 182.10, 582.10
Main references	(1) J. Essent. Oil Res., 2, p.91, (1990) (2) Perf.&Flav., 17(2), p.49, (1992) (3) Fenaroli, 1, p.264, (1995) (4) Food Chem. Toxicol. 27, p.11, (1989) (5) Mutat. Res., 326, p.199, (1995) (6) Xenobiotica, 17, p.1223, (1987); Food Chem. Toxicol., 26, p.87, (1988)
Data bases used	Chemical Abstracts, Toxline (1965-97) *Keywords:* Star anise, Illicium verum

Jasminum grandiflorum L.

CE No.	245
Steinmetz No.	601
FEMA No.	Jasmine absolute: 2598; Concrete: 2599; Oil: 2600; Spiritus: 2601
Order	Oleales
Family	Oleaceae
Name	E Spanish jasmine, royal jasmine
	F Jasmin
	D Jasmin
	I Jasmine
	SP Jazmin oloroso
Synonyms	J. officinale L. var. grandiflorum Bailey
Parts used	Flowers
Important constituents	Flowers: benzyl acetate (16.77%*), phenyl-acetic acid, linalool (3.3 %*), benzyl alcohol (0.73% *), methyl anthranilate (1.35%*), methyl heptenone, farnesol (7.93%), cis-3-hexenyl acetate, cis- and trans-linalool oxide, nerolidol, cis-jasmone, trans-methyl jasmonate, cis- and trans-ethyl jasmonates, jasmolactone, delta-jasmonic acid lactone, methyl dehydrojasmonate jasmonates (1). Benzyl benzoate (13.51%*), phytol (10.32%*), high paraffins (21.89%*) (4) * of jasmine absolute (4)
Active principles	Not known
Other chemical components	Not known
Products in which used	Jasmine absolute and concrete are used in alcoholic and non-alcoholic beverages, frozen dairy desserts, candy, baked goods, gelatin and puddings (1)
Level of use	Jasmine concrete/absolute: non-alcoholic beverages 2.84 ppm; ice-creams 3.01ppm; candy 4.22 ppm; baked goods 12.62 ppm; gelatin and puddings 3.02 ppm. Further, jasmine absolute is used in chewing gum (43.45 ppm) (2)
Preparation	Concrete, absolute essence and oil. Concrete is extracted from flowers using petroleum ether. Absolute is obtained by extraction of concrete with 95-96% ethanol. Oil is obtained by extraction of flowers with cold fat (natural flower oil), or from the absolute by steam distillation (essential oil) (1,2)
Main toxicological data	Benzyl acetate and benzyl benzoate were evaluated by JECFA (1987) and given an ADI of 5 mg/kg bw/d (5). Further, linalool has been evaluated by JECFA (1979) and given an ADI of 0.5 mg/kg

bw/d (6). Farnesol was negative in Ames test (5). No data on the toxic effects of J. grandiflorum flowers or any preparations thereof have been found

Data needed No data required

Specific observations None

Classification and limits **Concrete, absolute and essential oil: category 2**

National/int. evaluation Jasmine absolute, concrete, oil spiritus: CFR 182.20, 582.20; FEMA No..s 2598, 2599, 2600, 2601 respectively

Main references
(1) Leung (1980)
(2) Fenaroli Vol I (1994)
(3) Opdyke (1979)
(4) Anac, Flavour and fragrance journal (1) 115-119 (1986)
(5) JECFA 31st meeting (1987)
(6) JECFA 23rd meeting (1979)
(7) CoE Flavour. subst. and nat. sources of flavour. 4th ed.

Data bases used
Chemical Abstracts (1967-91)
Biosis (1973-91)
FSTA (1969-91)
Medline (1966-91)
Toxline (1965-91)
Keywords: Jasminum grandiflorum

Jasminum officinale L.

CE No.	246
Steinmetz No.	602
FEMA No.	Jasmine absolute: 2598; Concrete: 2599; Oil: 2600; Spiritus: 2601
Order	Oleales
Family	Oleaceae
Name	E White jasmine, common jasmine
	F Jasmin blanc
	D Jasmin, echter Jasmin
	I Gelsomino
	SP Jazmin morisco
Synonyms	-
Parts used	Flowers
Important constituents	Flowers: benzyl acetate, phenyl-acetic acid, linalool, benzyl alcohol, methyl anthranilate, methyl heptenone, farnesol, cis-3-hexenyl acetate, cis- and trans-linalool oxide, nerolidol, cis-jasmone, trans-methyl jasmonate, cis- and trans-ethyl jasmonates, jasmolactone, delta-jasmonic acid lactone, methyl dehydrojasmonate (1)
Active principles	Not known
Other chemical components	Not known
Products in which used	Jasmin absolute and concrete are used in alcoholic and non-alcoholic beverages, frozen dairy desserts, candy, baked goods, gelatin and puddings (1)
Level of use	Jasmine concrete/absolute: non-alcoholic beverages 2.84 ppm; ice-creams 3.01 ppm; candy 4.22 ppm; baked goods 12.62 ppm; gelatin and puddings 3.02 ppm. Further, jasmine absolute is used in chewing gum (43.45 ppm) (2)
Preparation	Concrete, absolute essence and oil. Concrete is extracted from flowers using petroleum ether. Absolute is obtained from concrete
Main toxicological data	Benzyl acetate and benzyl benzoate were evaluated by JECFA (1987) and given an ADI of 5 mg/kg bw/d (5). Linalool has been evaluated by JECFA (1979) and given an ADI of 0.5 mg/kg bw/d (6). Farnesol was negative in Ames test (7). 100 mg/kg of the jasmin-flower extract tested above produced progressive CNS depression, manifested as quiescence and long episodes of inactivity, 10-15 min after i.p. injection
Data needed	No data required

Specific observations	None
Classification and limits	Concrete, absolute and essential oil: category 2
National/int. evaluation	Jasmine absolute, concrete, oil spiritus: CFR 182.20, 582.20; FEMA No..s 2598, 2599, 2600, 2601 respectively
Main references	(1) Leung (1980) (2) Fenaroli Vol I (1994) (3) Int. J. Crude Drug Res. 26, 221-227 (1988) (4) Opdyke (1979) (5) JECFA 31st meeting (1987) (6) JECFA 23rd meeting (1979) (7) CoE Flavour. subst. and nat. sources of flavour. 4th ed.
Data bases used	Chemical Abstracts (1967-91) Biosis (1973-91) FSTA (1969-91) Medline (1966-91) Toxline (1965-91) *Keywords:* Jasminum officinale

Laurus nobilis L.

CE No.	255
Steinmetz No.	635
FEMA No.	Sweet bay leaves: 2124; Sweet bay leaves extract: 2613; Sweet bay leaves oil: 2125; Laurel berries: 2612
Order	Magnoliales
Family	Lauraceae
Name	E Laurel, sweet bay, bay laurel
	F Laurier
	D Lorbeer
	I Lauro nobile, alloro
	SP Laurel
Synonyms	-

Parts used	Leaves, berries
Important constituents	Leaves: isoquinoline alkaloids [mostly reticuline with some boldine, N-methylactinodaphnine, (+)-isidomesticine, (+)-neolitsine, actinodaphnine, nor-isodomesticine, launobine, nandigerine, and cryptodorine] (1) + sesquiterpenic lactones [mainly costunolide, with small amounts of artemorin, regnosin, santamarine and verlotorin] (2); + 1-3% essential oil + protein, lipid, carbohydrate, fiber, vitamines and minerals (3)
	Leaves essential oil: terpenic hydrocarbons [α-pinene 1-7.66%, a-thujene 0.06-0.45%, α-phellandrene 0.12-0.55%, sabinene 4.4-6.52%, β-pinene 1.81-4.36%, δ-3-carene 0.15-0.29%, myrcene 0.1-1.35%, p-cymene 0.06-0.70%, γ-terpinene 0.21-1.51%, terpinolene 0.16-2.33%, α-muurolene 0.22-0.89%]; oxygenated derivatives [eucalyptol 34-53%, linalool 11.57-18.73%, terpinen-4-ol 0.92-3.75%, geraniol 0.05-0.18%, α-terpineol 1.46-2.47%, borneol 0.18-0.47%, nerol 0.10-0.50%, spathulenol 0.10-0.28%, α-cadinol 0.38-1.12%, β-bisabolol 0.30-1.38%]; alkenyl benzenes [eugenol 1.44-2.03%, methyleugenol 1.73-11.8%]; esters [α-terpinyl acetate 9.54-18.02%, geranyl acetate 0.07-0.17%, linalyl acetate 0.29-0.94%, citronellyl acetate 0.09-0.24%, bornyl acetate 0.14-1.08%, methyl geranate 0.48-0.91%] (4,5,6,7)
	Berries: (30% pericarp, 70% seed): 20-34% fatty acids (= bay butter) [mainly lauric, palmitic, oleic and linoleic acids] (3)
Active principles	Methyleugenol
Other chemical components	Eucalyptol
Products in which used	Various foods

Level of use	Sweet bay leaves: baked goods 87.7 ppm, breakfast cereals 29 ppm, fats, oils 213.1 ppm, frozen dairy 5 ppm, fruit juice 560 ppm, meat products 768.9 ppm, processed vegetables 309.9 ppm, condiment, relish 1071 ppm, soft candy 5 ppm, soups 41.02 ppm, snack foods 220 ppm, non-alcoholic beverages 3 ppm, gravies 187.7 ppm (8) Sweet bay leaves extract: meat products and processed vegetables 20 ppm (8) Sweet bay oil: baked goods 27.81 ppm, frozen dairy 9.37 ppm, meat products 126.4 ppm, condiment, relish 214.9 ppm, soft candy 7.11 ppm, confection, frosting 50 ppm, gelatin, puddings 5 ppm, soups 0.4 ppm, non-alcoholic beverages 4.69 ppm, alcoholic beverages 7.41 ppm ppm, gravies 5.48 ppm ppm, chewing gum 1.44 ppm (8) Laurel berries: non-alcoholic beverages 450 ppm (8)
Preparation	Infusion, fluid extract, leaf essential oil
Main toxicological data	Eucaptol (1,8-cineole) see CE No. 185 Methyleugenol: not mutagenic with Ames test and with Eschrechia coli wp2 refersion test. Positive in the yeast assay, in the Bacillus subtilis DNA repair test, and in UDS. In mice treated with methyleugenol, presence of DNA adducts in liver. Hepatocarcinogenic effects occurred in pre-weaning mice treated with methyleugenol or with it's 1'hydroxy metabolite by i.p. injection Many metabolic pathways: allylic hydroxylation and oxidation of the side chain, O-demethylation, hydroxylation of the benzene ring. Ames test and UDS positive with 2,3-poxy-methyleugenol (7)
Data needed	No data required
Specific observations	Bay leaves used as a foodstuff
Classification and limits	**Leaves, leaves essential oil and extract: category 3 (with limits on eucalyptol and methyleugenol)** **Berries: category 2**
National/int. evaluation	Sweet bay leaves: CFR 182.10, 582.10 Sweet bay leaves extract: CFR 182.28, 582.20 Sweet bay leaves oil: CFR 182.20, 582.20 Laurel berries: CFR 182.20, 582.20 Laurel leaves extract: CFR 182.20, 582.20
Main references	(1) J. Nat. Prod., 45, p. 560, (1982) (2) J. Nat. Prod., 43, p.527, (1980) (3) Duke: Handbook of Medicinal Herbs, p.271, 7th Printing, (1989) (4) Acta Botanica Sinica, 32, p.878, (1990) (5) Nahrung, 36, p.494, (1992) (6) Perf. Flav., 15(3), p.67, (1990); 18(3), p.65, (1993) (7) Flavour Fragr. J., 12, p.91, (1997) (8) Fenaroli, 1, p.163, (1995)
Data bases used	Chemical Abstracts, Toxline (1965-97) *Keywords:* Laurel, Sweet bay, Laurus nobilis

Mangifera indica L.

CE No.	270
Steinmetz No.	692
FEMA No.	-
Order	Sapindales
Family	Anacardiaceae
Name	E Mango tree
	F Manguier
	D Mangobaum
	I Mango
	SP Mango
Synonyms	
Parts used	Fruit, peel of fruit
Important constituents	Fruit: water, sugars, amino-acids, beta carotene, vitamins A and C and volatile compounds [terpenic hydrocarbons mainly α-terpinolene, car-3-ene, α-terpinene, limonene, β-phellandrene, myrcene, α-pinene, α-gurjumene, caryophyllene, α-acetate; + alcohols: hexa-1-ol, *cis*-hex-3-en 1-ol, 2-methylbutan-1-ol; + carbonyl compounds: hexanal, furfural, 2,5-dimethyl-4-hydroxy-3(2H)-furanone; + lactones mainly γ-octalactone, γ-butyrolactone, χ-nonalactone and γ-decalactone] (1,2,3,4,5) Peel of fruit: mainly 5-substituted resorcinols: 5-(12-*cis*-heptadecenyl)resorcinol and 5-pentadecylresorcinol (6)
Active principles	Not known
Other chemical components	Not known
Products in which used	Mango is a foodstuff consumed as such
Level of use	No information available
Preparation	Juice and mashed fruit
Main toxicological data	No relevant data found
Data needed	No data required
Specific observations	Fruit is used as a foodstuff
Classification and limits	Fruit: category 1 Rind: category 2
National/int. evaluation	UK FACC 1976, Appendix 1 (fruit), Appendix 2 (rind)

Main references   (1) Phytochemistry 27:2189 (1988)
(2) J. Agr. Food Chem. 36:137 (1988)
(3) J. Agr. Food Chem. 38:1556 (1990)
(4) J. Food Science 52:353 (1987)
(5) Lebensm. Wiss. u. Tchnol. 25:374 (1992)
(6) Phytochemistry 25:1093 (1986)

Data bases used   Chemical Abstracts, Pascal (1965-93)
Keywords: Mango

Marsdenia cundurango Rchb.f.

CE No.	272
Steinmetz No.	-
FEMA No.	-
Order	Gentianales
Family	Asclepiadaceae
Name	E Condurango, eagle-vine, condor plant F Condurango D Kondurangostrauch, Kondorliane, Geierpflanze I Condurango SP Codurango
Synonyms	Marsdenia condurango Reichb.f., M. reichenbachii Triana, Gonolobus condurango (Reichb, f.), Pseudomarsdenia condurango (Reichb, f.) Schlechter
Parts used	Stem bark (1)
Important constituents	1-3% Condurangine (various bitter condurango glycosides i.e. esters of polyhydroxylated pregnane glycosides, mainly conduran-goglycosid A consisting of kondurangogenin A, cymarose, olean-drose, 6-desoxy-3-O-methylallose and glucose), other polyhydroxylated pregnane derivatives such as sarcostin, drevogenine-D, dihydro-drevogenine-D, marsdenin; the pseudoalkaloids condurangamin A and B (ca. 0.001%) and gagaminin (< 0.015%); the phenolic compounds chlorogenic acid, caffeic acid, cinnamic acid, p-coumaric acid, vanillin (0.022-0.045%), the coumarins coumarin, umbelliferon, aesculetin, cichoriin; the flavonoid glycosides (0.012-0.084%) trifolin, hyperin, quercitrin, rutoside, saponarin; various cyclites (ca. 0.75%; conduritol 0.3-0.5% and others) (1,2,3). Furthermore digitaloids and esterglycosides, vincetoxin, sitosterin and β-amyrin as cinnamates, ca. 0.01% essential oil, ca. 6% rubber (4)
Active principles	Coumarin
Other chemical components	Not known
Products in which used	Bark is used in non-alcoholic beverages, alcoholic beverages (bitter wines (1)) and candies
Level of use	Bark extract: non-alcoholic beverages 10-100 ppm, alcoholic beverages 500 ppm, candies 10-100 ppm (IOFI 1998). Bark: traditional use of 50-100 g bark per liter of wine
Preparation	Extracts, fluid extracts, tinctures (1)

Main toxicological data	Condurangin was toxic in animals after i.v., s.c. and p.o. application (hypersalivation, vomiting, ataxia, etc. leading to death through respiratory paralysis) (5,6,7), lethal dose of condurangin for dogs and cats 40-45 mg/kg bw (p.o.), 30 mg/kg bw (s.c.) and 20 mg/kg bw (i.v.) (8,9). An antitumor activity in mice has been reported for condurangoglycosides A_0 and C_0 (8), which were found to be differentiation inducers in vitro (10)
Data needed	No data required
Specific observations	None
Classification and limits	**Bark and preparations: category 4 (with limits on coumarin)**
National/int. evaluation	None
Main references	(1) Wichtl M. Teedrogen. 2nd Ed., WVG, Stuttgart (1989) (2) Zeitschrift für Phytotherapie, 8, 38-41 (1987) (3) Hager's Handbuch der Pharmazeutischen Praxis, 5th Ed., Haenseler R. et al. (Ed.), Springer Verlag, Berlin (1990) (4) Hoppe H.A., Drogenkunde, 8th Ed., de Gruyter, Berlin (1975) (5) Helv. Chim. Acta., 53, 221-245 (1970) (6) Frohne D., In: DeSmet et al. (Ed.) Adverse Effects of Herbal Drugs, Springer, Berlin, Vol. 1, 157-160 (1992) (7) Planta Med. 10, 107-117 (1962) (8) Chem. Pharm. Bull., 28, 1954-58 (1980) (9) Chem. Pharm. Bull., 29, 2725-2730 (1981) (10) Chem. Pharm. Bull., 42, 611-616 (1994)
Data bases used	Medline (1966-98) Embase (1980-98) Biological Abstracts (1989-98) CC Life (2/97-2/98) *Keywords:* Marsdenia condurango, eagle-vine, condurango

Murraya koenigii (L.) Spreng.

CE No.	2076
Steinmetz No.	-
FEMA No.	-
Order	Rutales
Family	Rutaceae
Name	E Indian curry leaf tree
	F -
	D Indischer Curryblätterbaum
	I -
	SP
Synonyms	-
Parts used	Leaves
Important constituents	Indian curry leaves: carbazole alkaloids (qualitative composition): mainly: koenigicine, koenimbine, cyclomahanimbine, mahanimbine, girinimbine, isomahanimbine, koenimbine, mahanimbidine, mahanine, koenine, koenigine, koenidine, mukonicine, mahanimbicine; coumarinic derivative: scopolin (1, 2, 3); and volatile compounds
	Volatile compounds of indian curry leaves (essential oil): terpenic hydrocarbons [β-thujene 4.3%; β-caryophyllene 28.7%; β-elemene 6.8%; β-phellandrene 6.1%; limonene 2.1%; *trans* β-ocimene 1.9%; β-gurjunene 21.4%; β-bisabolene 2.8%; A selinene 12.5%; g-cadinene 2.5%; α-selinene 2.9%; α-copaene 0.9%; β-pinene 0.7%] (4)
	Volatile compounds of malaysian curry leaves: terpenic hydrocarbons [α-pinene 17.5%; α-thujene 1.6%; camphene 0.2%: β-pinene 3.7%; sabinene 4.1%; myrcene 2.3%; α-phellandrene 4.8%; α-terpinene 2.8%; limonene 5.1%; β-phellandrene 24.4%; (*E*)-β-ocimene 1.8%; γ-terpinene 4.9%; π-cymene 1%; terpinolene 1.1%; β-caryophyllene 7.3%; α-humulene 0.6%; δ-cadinene 0.6%; linalool 0.6%; *cis*-sabinene hydrate 0.5%; *trans*- sabinene hydrate 0.4%]; terpenic alcohols and ester [terpinen-4-ol 6.1%; α-terpineol 1.6%; lavandulyl acetate 0.9%] (5)
Active principles	Not known
Other chemical components	Not known
Products in which used	Leaves are commonly added to curry; tincture in beverages and candy

Level of use	Tincture: beverage 20 ppm; candy 100 ppm (IOFI, 41st Meeting, October 1997
Preparation	Tincture, essential oil
Main toxicological data	No relevant data found
Data needed	No data required
Specific observations	None
Classification and limits	**Leaves and preparations: category 2**
National/int. evaluation	None
Main references	(1) Fortschritte der Chemie organischer Naturstoffe, 34, p.299, (1977) (2) Medicinal Plants of India, Indian Council of Medical Research, New Delhi, Vol 2, p.289, (1987) (3) Fitoterapia, 65, p. 49, (1994) (4) Phytochemistry, 21, p.1653, (1982) (5) J.Essent. Oil Res., 5, p.371, (1993)
Data bases used	Chemical Abstracts: 1965-96 *Keywords:* Murraya koenigii

Musa L. species, Musa sapientum L.

CE No.	294
Steinmetz No.	736
FEMA No.	-
Order	Zingiberales
Family	Musaceae
Name	E Banana (M. sapientum)
	F Banane, Bananier
	D Banane
	I Banano
	SP Platano
Synonyms	Banana = M. paradisiaca L. var sapientum; Cavendish Banana = Chinese Banana = Dwarf Banana (M.nana Lour.); Plantain = Musa X paradisicica) (1,2)

Parts used	Fruit (edible part) as food
Important constituents	In fruit: isoamyl acetate 0.2-25 ppm; *trans*-2-hexenal 18-76ppm; 2-pentanone 27ppm; ethyl acetate 0-169ppm (3)
Active principles	Not known
Other chemical components	Not known
Products in which used	Used as food, also used as flavouring in beverages and many food types
Level of use	No information available
Preparation	Extract, juice, spray dried powder
Main toxicological data	No data on plant. Isoamyl acetate – no evidence of any effect when tested for induction of mitotic chromosomal malsegregation, mitotic recombination and point mutation in a diploid yeast strain D61.M (4). All principal components of fruit listed as category A Blue Book 4th Ed. Volume I
Data needed	No data required
Specific observations	None
Classification and limits	Fruit: category 1
National/int. evaluation	UK FACC (1976), Fruit Appendix 1

Main references	(1) Usher (1974) (2) TDRI G201 Selected European markets for speciality and tropical fruit and vegetables. Corrine Joy 1987 ODA Crown Copywright pp24 and 62 (3) TNO-CIVO Institute, Volatile compounds in Food, Vol. 1, 1982 (4) Mutat Res 149, p339-351 (1985)
Data bases used	Biosis (1973-90) Chemical Abstracts (1967-90) FSTA (1969-90) Medline (1966-90) Embase (1982-90) *Keywords:* Musa species, Banana, M acuminata, M. nana, M. paradisiaca, M sapientum, Plantain, Red Banana, Apple Banana, Green B

Olea europaea L.

CE No.	309
Steinmetz No.	769
FEMA No.	-
Order	Oleales
Family	Oleaceae
Name	E Olive tree
	F Olivier
	D Ölbaum, Olivenbaum
	I Oliva
	SP Olivo
Synonyms	Olea sativa Hoffsmeg et Link
Parts used	Fruits, leaves
Important constituents	Leaves: glucosides (oleuropeine, 0.75%), luteolin, olivin, triterpenes (oleanoic acid (2-3%) and homoolestranol), elenolide, oleastearol, malic acid, tartaric acid, glycolic acid, lactic acidoleanolic acid (2-3%) (1,3), alkaloids (cinchonidine, cinchonine, dihydrocinchonine; together ca. 0.008%)(4) Fruits: olive oil (50%), oleuropein, demethyloleuroein, cyanidin-3-glucosid, 3-rutinosid (3). Fruit kernels: estrone (2)
Active principles	Not known
Other chemical components	Not known
Products in which used	Olive oil in food, leaf extract in beverages and for medical use
Level of use	Annual use of leaves in Europe 200 kg. Use level of leaves 0.2 g/l in beverages (IOFI)
Preparation	Olive oil, leaf extract
Main toxicological data	Oleuropein has hypotensive action due to vasodilatation of vessels. It also relaxes smooth muscles (3). I.p. injection of 100-1000 mg oleuropein/kg body weight in mouse did not cause any mortality or toxicological lesions during the 7 days when the animals were observed (5)
Data needed	Quantitative chemical data and, if necessary, 28-day oral study and mutagenicity studies on relevant leaf extract
Specific observations	None
Classification and limits	**Fruit: category 1** **Leaves: category 5**

National/int. evaluation	None
Main references	(1) Camurati et al., La vista italiana delle sostanze grasse vol. LVIII, p541-547. 1981 (2) Amin & Bassiouny, Phytochem. 18, 344. 1979 (3) List & Hörhammer 1968-80 (4) Schneider & Kleinert, Planta medica 22 (2), 109, 1972 (5) Petkov & Manolov, Arzneim.-Forsch., 22, 1476, 1972
Data bases used	Chemical Abstracts (1967-97) Biosis (1973-97) FSTA (1969-88) Medline (1966-97) Toxline (1981-88) *Keywords:* Olea europaea

Piper cubeba L.

CE No.	345
Steinmetz No.	849
FEMA No.	Cubeb: 2338; Cubeb oil: 2339
Order	Piperales
Family	Piperaceae
Name	E Cubeb
	F Cubèb
	D Kubebenpfeffer, Schwanzpfeffer, Stielpfeffer
	I Cubebe
	SP Cubeba
Synonyms	Cubeba officinalis Miq.
Parts used	Fruits
Important constituents	Fruits contain 10-20% volatile oil, 2.5% cubebin, 1-1.7% amorphous cubebic acid, resins, gum, lignans (4) and fat. *Cubeb oil* contains 17% sesquiterpenes (caryophyllene, cadinene, (14% in oil), alpha- and beta-cubebene, copaene, and 1-isopropyl-4-methylene-7-methyl-1,2,3,6,7,8,9-heptahydronaphtalene) and 9.5% monoterpenes (sabinene, (5-33% in oil), alpha-thujene, (1,5% in oil) beta-phellandrene, alpha-pinene, traces of myrcene, p-cymene, terpinolene, beta-pinene, alpha-phellandrene, gamma- and alpha-terpinene, limonene, and ocimene). Further, oxygenated terpenes (1,4-cineole, alpha-terpineol, cadinol, and cubebol (1,5)
Active principles	Not known
Other chemical components	Not known
Products in which used	Alcoholic and non-alcoholic beverages, ice-cream, candy, baked goods, condiments and meat
Level of use	Dried fruits: 850 ppm in non-alcoholic beverages Cubeb oil: non-alcoholic beverages 7.8 ppm, ice-cream 17.2 ppm, candy 18.2 ppm, baked goods 16.5 ppm, condiments 38.2 ppm, meats 102.1 ppm, alcoholic beverages 30.0 ppm, gelatin and puddings 14.4 ppm (2)
Preparation	Cubeb oil (2)
Main toxicological data	No relevant data found
Data needed	28-day oral study and study on mutagenicity on relevant extracts
Specific observations	None

Classification and limits	Fruit and oil: category 5
National/int. evaluation	Approved for food use (FDA §172.510) Cubeb CFR 172.510; Cubeb oil CFR 172.510
Main references	(1) Leung 1980 (2) Fenaroli 1995 (3) Tyler, 1976 (4) Phytochem. 25:2, 487-489, 1986
Data bases used	Chemical Abstracts (1967-91) Biosis (1973-91) FSTA (1969-91) Medline (1966-91) Toxline (1981-91) *Keywords:* Piper cubeba

Plantago lanceolata L.

CE No.	352
Steinmetz No.	861
FEMA No.	-
Order	Plantaginales
Family	Plantaginaceae
Name	E ribgrass, ribwort plantain, buckhorn plantain
	F plantain, lancéolé
	D Spitzwegerich
	I Arnoglossa. Plantagginé pettiti
	SP Llanten menor
Synonyms	-

Parts used	Leaves (herb)
Important constituents	Aucubin (0.5-1.8%)(1)
Active principles	Not known
Other chemical components	Not known
Products in which used	Leaves in salads as a possible emergency food (2). Beverages, foods (IOFI 1994)
Level of use	Annual use of leaves in Europe 900 kg. Use levels of leaves in beverages 0.001 g/l; in food 0.002 g/kg (IOFI 1994)
Preparation	None
Main toxicological data	No relevant data found. Alcoholic extract from P. lanceolata show liver-protective activity in mice intoxicated with CCl_4. This effect is ascribed to aucubin, which acts inhibitory on RNA synthesis. The toxicity of aucubin is, however, relatively low. Administration of 80 mg/kg, 4 times a week, in mice did not affect serum enzyme levels or chemical parameters. Minimal lethal dose for mice was 0.9 g (3)
Data needed	Chemical composition and, if necessary, 28-day study and studies on mutagenicity on leaves
Specific observations	None
Classification and limits	Leaves: category 5
National/int. evaluation	None

Main references
(1) Juneby, Junebys medicinalväxter, Reformförlaget, Malmö, Sweden (1984)
(2) Usher, A Dictionary of Plants used by Man, Constable, London (1974)
(3) Chang & Yun, Adv. in Chinese med. mater. res. World Scient. publ. Co, Singapore. p. 269-285. 1985

Data bases used
Chemical Abstracts (1973-97)
FSTA (1969-97)
Medline (1966-87)
Toxline (1981-97)
Keywords: Plantago lanceolata, acubin, plantagin

Populus nigra L.

CE No.	361
Steinmetz No.	894
FEMA No.	-
Order	Salicales
Family	Salicaceae
Name	E Black poplar, Lombardy poplar F Peuplier noir D Schwarzpappel I Pioppo nero SP Chopo
Synonyms	Populus pannonica Kit., populus nigra L. var. italica Lombardy poplar
Parts used	Leaf buds, bark
Important constituents	Bud essential oil: (0.5% bud): β-caryophyllene (= α-humulene in hops), cis-3-hexen-1-ol (2% & 16.7% in two samples), 1,2-cyclohexanedione (14% & 41.3% in 2 samples), eugenol, (28.7% & 13.2% in 2 samples), 1-octadecanol (16.7% & 13.2% in 2 samples). Bud oleoresin: Salicin, populin, gallic acid, chrysin, tannins. Buds: flavonoids including quercetin, kaempferol and galangin; phenolic acids including-dimethylallylcaffeic acid. Poplar buds (species unspecified): cinnamoyl cinnamate (styracin) major component of fragrance) (1,2,3)
Active principles	Not known
Other chemical components	Not known
Products in which used	Liqueurs (1). Sweets. Note: (4) does not list uses, but under Regulatory status refers to US legislation: CFR 172.510 (in alcoholic beverages only). Neither (1) nor (4) mentions use in food, only pharmacology, perfumery and alcoholic drinks
Level of use	Buds: alcoholic beverages 400 ppm; sweets 10 – 400 ppm; food 40 ppm
Preparation	Essential oil (from buds or oleoresin), oleoresin (or "concrete"), tincture, fluid extract, soft extract, infusion
Main toxicological data	Bark/buds: no toxicity data found. Individual components: 1,2-cyclohexanedione (90d study, rats), NOEL: 5.8 mg/kg (5). Salicin, gallic acid mildly irritating (Merck index). Quercetin, kaempferol and galangin mutagenic (6). Quercetin carcinogenicity data inconclusive (6) 8,8-Dimethylallyl caffeic acid contact allergen (7) β-caryophyl-

	lene category B; *cis*-3-hexen-1-ol, eugenol category A according to Blue Book 4th Ed. Volume I
Data needed	Data on the importance of Populus nigra as source for poplar bud oil. Mutagenicity and 28-day oral studies on essential oil. Further chemical/toxicity/usage data, chemical composition and, if necessary, toxicity data on bark
Specific observations	None
Classification and limits	**Buds: category 5** **Bark: category 5**
National/int. evaluation	UK FACC (1976) Appendix 2 FDA 172.510 Poplar buds (Populus balsaminifera L., P. candicans Ait. or P. nigra L.) in alcoholic beverages only
Main references	(1) Fenaroli (1975) (2) Chem. Abs. 097 (08) 060805 (1982) (3) Phytochem. 8: p2425-2426 (1969) (4) Fenaroli (1995) (5) Food and Drug Res Labs Rep (Compound 14841) (1974) (6) Prog. clin. Biol. Res. 206: p33-43 (1986) (7) Z. Naturforsch 42c p1030-34 (1987) (8) Belgian J. Bot. 129: 123-130 (1997)
Data bases used	Biosis (1973-97) Chemical Abstracts (1967-97) Toxlit (1981-97) Medline (1966-90) Embase (1974-97) *Keywords:* Populus nigra, Black Poplar

Quercus alba L.

CE No.	388
Steinmetz No.	934
FEMA No.	Chips extract: 2794
Order	Fagales
Family	Fagaceae
Name	E White oak
	F Chêne blanc
	D Weisseiche
	I Quericia
	SP Roble
Synonyms	-
Parts used	Wood, bark
Important constituents	In wood: tannins (gallic acid, ellagic acid, tannic acid, ellagotannic acid), coniferylaldehyde, tetracosylferulat, sterols (β-sitosterol, stigmasterol, campesterol). In bark: tannins, resin, carbohydrates (pectin, laevulin, quercitol). (2)
Active principles	Not known
Other chemical components	Not known
Products in which used	Oak chips extract is used in non-alcoholic beverages, alcoholic beverages, ice-cream, candy, baked goods. Oak is used in vermouth flavours (1)
Level of use	Non-alcoholic beverages 8.66 ppm; alcoholic beverages 43.55 ppm; ice-cream etc. 88.64 ppm; candy 90.00 ppm, baked goods 90.00 ppm (1). Annual use in Europe 21300 kg wood (IOFI). Use levels in alcoholic beverages 3g/l (IOFI 1994). Use levels in beverages: bark 0.2-5 g/l; bark extract 100-500 ppm; wood 0.2-10 g/l; wood extract 10-700 ppm (IOFI 1994)
Preparation	Wood chips, wood extract, bark extract
Main toxicological data	No relevant data found
Data needed	Quantitative data on chemical composition of wood and bark extracts and, if necessary, 28-day oral study and studies on mutagenicity
Specific observations	None
Classification and limits	**Wood chips: category 2**
	Bark and extracts: category 5

National/int. evaluation Oak chips extract: CFR 172.510 (1)

Main references (1) Fenaroli Vol I (1995)
(2) List & Hörhammer (1967-80)

Data bases used FSTA (1969-90)
Biosis (1973-90)
Medline (1966-90)
Keywords: Quercus, white oak

Saccharum officinarum L.

CE No.	2100
Steinmetz No.	-
FEMA No.	-
Order	Graminales
Family	Gramineae
Name	E Cane sugar
	F Canne à sucre
	D Zuckerrohr
	I Canna di zucchero
	SP
Synonyms	-
Parts used	Stems
Important constituents	Klasonlignin (16.4%, of which 16.3% total aldehyde, 5.3% vanillin, 9.0% syringaaldehyde) (1). About 12-20% saccharose (located in the juice of the stem) (2). Polysaccharides (ca. 2.5%) (3). About 0.1-0.25% wax (on stem surface) (2) containing policosanol (a defined mixture of 8 higher aliphatic alcohols with octacosanol as main component) (4)
Active principles	Not known
Other chemical components	Not known
Products in which used	No information available
Level of use	No information available
Preparation	Molasses extract (by product of sugar-refining process: the syrup or the "mother-water" that is separated from the grains of raw sugar in the process of manufacture (molasses, molasses concentrate, molasses extract) (5)
Main toxicological data	No relevant data found
Data needed	No data required
Specific observations	Sugar cane and molasses are used as a foodstuff
Classification and limits	**Cane and molasses: category 1**
National/int. evaluation	Regulatory status in USA: molasses, molasses concentrate, molasses extract: CFR 182.20, 582.20 (5)

Main references

(1) Hegnauer R., Chemotaxonomie der Pflanzen, Vol. 2, Birkhäuser, Basel (1963-)
(2) Hoppe H.A., Drogenkunde, 8th Ed., de Gruyter, Berlin (1975)
(3) J. Ethnopharmacology, 14, 261-268 (1985)
(4) Toxicol. Lett. 73, 81-90 (1994)
(5) Fenaroli (1995)

Data bases used

Medline (1966-98)
Embase (1980-98)
Biological Abstracts (1989-98)
CC Life (2/97-2/98)
Keywords: Saccharum officinarum, cane sugar

Sambucus nigra L.

CE No.	417
Steinmetz No.	1016
FEMA No.	Flowers: 2406
Order	Dipsacales
Family	Caprifoliaceae
Name	E Black elder
	F Sureau noir
	D Schwarzer Holunder
	I Sambuco nero
	SP Sauco
Synonyms	Sambucus vulgaris Neck

Parts used	Fruits, flowers, flower tips, leaves
Important constituents	Flowers: Essential oil contains free fatty acids and alkanes, triterpenes, ursolic acid, sterols, flavonoids and flavone glucosides (quercetin, kaempferol, isoquercitrin and rutin). Traces of sambunigrin
	Fruits: rutin, isoquercitrin, sambucin,(cyanidin-3-rhamnoglucosid), anthocyanglucosides. Seeds contain sambunigrin (3). The reputed content of sambunigrin in unripe fruit is not confirmed in any of the references used
	Leaves: derivatives contain sambunigrin, a mandelonitrile glucoside
Active principles	Hydrocyanic acid. Seeds and leaves contain sambunigrin, a mandelonitrile glycoside, which liberates HCN on hydrolysis
Other chemical components	Not known
Products in which used	Flowers are used in alcoholic (bitters and vermouths) and non-alcoholic beverages, frozen dairy desserts, candy, baked goods, gelatin and puddings. (1,4) Extract from leaves used in alcoholic drinks only (4)
Level of use	Flowers: Baked goods 30.0 ppm; frozen dairy 44.0 ppm; soft candy 30.0 ppm; gelatin, puddings 30.00 ppm; non-alcoholic beverages 489.5 ppm; alcoholic beverages 14.50 ppm (4). Use level of flowers in alcoholic beverages 3g/l (IOFI 1994)
Preparation	Fluid extract, tincture (20% in 60% ethanol or 15% in 55% ethanol) and crude
Main toxicological data	The roots and stem of elder have caused poisoning in man, and have a reputation as purgatives lax. Berries eaten raw can cause

	nausea and vomiting. Both berries and flower, however, have long been used, apparently safely, after cooking (2)
Data needed	Chemical composition and, if necessary 28-day oral study and mutagenicity studies on extract of leaves
Specific observations	None
Classification and limits	**Fruit: category 3 (with limits on hydrocyanic acid)** **Flowers and flower tips: category 1** **Leaves and extracts: category 5 (with limits on hydrocyanic acid)**
National/int. evaluation	Elder flowers: CFR 182.10, FEMA No.. 2406 Elder flowers extract: CFR 182.20; 582.20 Elder tree leaves extract 172.510 (in alcoholic beverages only, not to exceed 25 ppm prussic acid in the flavour)
Main references	(1) Leung (1980) (2) Duke (1985) (3) List & Hörhammer (1968-89) (4) Fenaroli (1995)
Data bases used	Chemical Abstracts (1967-97) Biosis (1973-897) FSTA (1969-88) Medline (1966-97) *Keywords:* Latin and English names, rutin

Santalum album L.

CE No.	420
Steinmetz No.	1022
FEMA No.	Wood oil: 3005
Order	Santalales
Family	Santalaceae
Name	E Sandaltree, White Sandaltree
	F Santal blanc, Santal jaune
	D Gelbsandelbaum, Santalbaum
	I Sandalo bianco
	SP Sandalo
Synonyms	Sirium myrtifolium L.
Parts used	Wood (1, 2)
Important constituents	East Indian Sandalwood oil: 90% is α-santalol. 6% is sesquiterpene hydrocarbons (α- and β-santalenes, epi-β-santalene, α- and β-curcumenes), possibly β-farnesene, dendrolasin. (1,2,3,4)
Active principles	Not known
Other chemical components	Not known
Products in which used	Oil used in baked goods, frozen dairy products, soft candy, gelatin, puddings, alcoholic and non-alcoholic beverages, hard candy and chewing gum (4), ginger ale, and in floral, fruit, honey and spice flavourings (2)
Level of use	Oil: baked goods 9.72ppm; frozen dairy products 3.63 ppm; soft candy 9.71 ppm; gelatin, puddings 0.73 ppm; alcoholic beverages 0.77 ppm; non-alcoholic beverages 1.96 ppm; hard candy 89.98 ppm; chewing gum 2.56 ppm (4)
Preparation	East Indian Sandalwood Oil stem distilled from wood (5), also called Oil of Santal (6). Oleoresin also used
Main toxicological data	Irritant dose (dermal, rabbit) 500 mg/24hr (7). Overdose (amount not specified) produces gastric and renal irritation and skin eruption (2)
	α and β santalol category B, Blue Book, 4th Ed. Volume I – request for 28-day study
Data needed	28-day oral study and mutagenicity study on sandalwood oil and/or 28-day study on the α and β santalol. Quantitative chemical and toxicological data on flavourings produced from oleoresins
Specific observations	None

Classification and limits	**Wood: category 5**
National/int. evaluation	UK FACC (1976) Appendix 2
	USA FDA 172.510
	Food Chemical Codex III p268
Main references	(1) Duke (1985)
	(2) MAFF (1995)
	(3) Helva Chim. Acta 59, 3, 737-749 (1976)
	(4) Fenaroli (1995)
	(5) Arctander (1960)
	(6) Usher (1974)
	(7) Niosh. Reg. Tox Effect Chem. Sub. 2, 207 (1979)
	(8) Frag. Raw Mater. Monographs p 989
Data bases used	Biosis (1973-89)
	Chemical Abstracts (1967-97)
	FSTA (1969-89)
	FROSTI (1975-89)
	Toxlit and Toxnet (1981-97)
	Medline (1966-97)
	Embase (1974-89)
	Keywords: Santalum album, Sandalwood, White Sandalwood, East Indian Sandalwood

Schinus molle L.

CE No.	427
Steinmetz No.	1034
FEMA No.	-
Order	Sapindales
Family	Anacardiaceae
Name	E American pepper tree
	F Poivrier d'Amérique, Faux poivrier
	D Mollebaum, Schinuspfefferbaum, falscher Pfefferbaum
	I Schino, Albero di pepé di Peru, Falso pepé
	SP Pimentero falso
Synonyms	-
Parts used	Fruit, leaf
Important constituents	Essential oil of fruit: terpenic hydrocarbons [limonene 4.0-9.0% a-phellandrene 5.3-26.0%; β-phellandrene 4.8-7.2%; myrcene 5.0-19.0%; α-pinene 1.4-3.1%; δ-cadinene 4.7-6.5%; π-cymene 6.2-11.5%]; phenolic compound = carvacrol 0.6-1%; ester = methyl octanoate 0.2-1.6% (1)
	Essential oil from leaf: monoterpenic and sesquiterpenic hydrocarbons: [β-pinene 13.95%; sabinene 12.92%; myrcene 5.46%; limonene 0.68%; β-phellandrene 0.30%; γ-terpinene 1.13%; β-caryophyllene 7.68%; α-humulene 0.57%; germacrene-D 12.08%; bicyclogermacrene 29.20%; δ-cadinene 1.26%]; oxygenated terpenic compounds [linalool 0.71%; terpinen-4-ol 10.57%; α-terpineol 1.25%; carophyllene oxide 0.53%; germacrone 0.75%] (2)
Active principles	Not known
Other chemical components	Carvacrol
Products in which used	Baked goods; frozen dairy; meat products; condiment, relish; soft candy; gelatin, puddings; alcoholic and non-alcoholic beverages
Level of use	Essential oil of fruit: baked goods 18.79 ppm; frozen dairy 34.52 ppm; meat products 200.8 ppm; condiment, relish 55.48 ppm; soft candy 14.38 ppm; gelatin, puddings 19.18 ppm; non-alcoholic beverages 11.51 ppm; alcoholic beverages 25 ppm (3)
	Leaves: 3000 ppm in beverages and foods (IOFI)
Preparation	Essential oil, fluid extract, tincture
Main toxicological data	No relevant data found
Data needed	No data required

Specific observations	None
Classification and limits	**Fruit: category 4 (with limits on carvacrol)** **Leaves, essential oil of fruit and leaves: category 2**
National/int. evaluation	None
Main references	(1) Perf. Flav., 9(5), p.65, (1984) (2) J. Essent. Oil Res., 8, p.71, (1996) (3) Fenaroli, 1, p.257, (1995)
Data bases used	Chemical Abstracts – Pascal: 1956 – June 1996 *Keywords:* Schinus molle

Swertia chirata Buch.-Ham. ex Wall.

CE No.	440
Steinmetz No.	-
FEMA No.	-
Order	Gentianales
Family	Gentianaceae
Name	E Chirata, chirayta, bitter stick
	F Chirette indien
	D Tarant, Indische Chiretta
	I Chiretta
	SP -
Synonyms	Swertia chirata (Roxb.) Buch.-Ham., Swertia indica, Agathodes chirayata D. Don, Gentiana chirayata Roxb., Ophelia chirayata Griseb
Parts used	Whole plant with small portion of the root (1)
Important constituents	Bitter secoiridoid glycosides such as swertiamarin (ca. 0.4%), gentiopikroside (trace amounts to 0.45% in crude drug (2), and sweroside, and the much bitterer biphenylcarboxylic acid esters of sweroside and swertiamarin: amarogentin (chiratin; 0.054-0.56% in crude drug) and amaroswerin (trace amounts to 0.25% in crude drug) (2,3); numerous tetraoxygenated xanthones including bellidifolin (1,5,8-trihydroxy-3-methoxyxanthone), methylbellidifolin (swerchirin; 1,8-dihydroxy-3,5-dimethoxyxanthone), desmethylbellidifolin (1,3,5,8-tetrahydroxy-xanthone), 5,8-dimethylbellidifolin (1-hydroxy-3,5,8-trimethoxyxanthone), methylswertianin (1,8-dihydroxy-3,7-dimethoxyxanthone), norswertianin (1,3,7,8-tetrahydroxyxanthone), decussatin (1-hydroxy-3,7,8-trimethoxyxanthone), 1,3,8-trihydroxy-5-methoxyxanthone (4,5), and chiratol (1,5-dihydroxy-3,8-dimethoxyxanthone) (6), the xanthone glycoside mangiferin (0.12% of whole plant; 1,3,6,7-tetrahydroxyxanthone-C2-β-d-glucoside) (7), and several others (8); triterpenes (swertane triterpenoids swertanone, swertenol, episwertenol, swerta-7,9(11)-dien-3β-ol and pichierenol; chiratenol, gammacer-16-en-3β-ol, β-amyrin, ψ-taraxasterol, lupeol and erythrodiol (9) and kairatenol); alkaloids (gentianine, gentiocrucine, enicoflavine (10)) (6,11). Also ophelic acid (1)
Active principles	Xanthones
Other chemical components	Not known
Products in which used	Alcoholic beverages (bitter tonics) (1)

Level of use	Herb tincture: non-alcoholic beverages 50 ppm, alcoholic beverages 400-1530 ppm (IOFI 1998). Average maximum use levels of herb in alcoholic beverages 0.0016% and in non-alcoholic beverages 0.0008% (11)
Preparation	Tincture (20% in 20 to 40% ethanol) and fluid extract (1)
Main toxicological data	Some constituents have pharmacological properties such as antituberculous (swertianin, norswertianin, methylswertianin), hepatoprotective (amarogentin) and antimalarial activity (swerchirin) (11). Amarogentin also inhibits the catalytic activity of DNA topoisomerase I of Leishmania donovani in vitro (12) Total xanthones extracted from the aerial parts have anti-inflammatory activity in rats at oral doses of 50 mg/kg bw (13) The dried ethanolic extract has been shown to have a protective effect against gastric ulcers in rats (14) Swertiamarin shows a CNS-depressant effect in mice and rats at doses of 50-100 mg/kg bw i.p., reversing the CNS-stimulating effect of mangiferin administered at doses of 100 mg/kg bw i.p. (7). A mutagenic activity resulting in a positive Ames Test with metabolic activation using S9-mix due to the content in xanthones has been shown in vitro with the methanolic extract (3,4,15). The mutagenic principles are represented by seven tetraoxygenated xanthones detected in the methanolic extract: methylbellidifolin, methylswertianin, swertianin, bellidifolin, norswertianin, desmethylbellidifolin, 5,8-dimethylbellidifolin. About 80-90% of the mutagenicity of the methanolic extract is accounted for by the sum of the activity of the seven xanthones with bellidifolin being the most active (15). After treatment with nitrite, the methanolic extract showed mutagenic activity on Salmonella typhimurium TA98 and 100 also in the absence of S9-mix due to the acyl moiety (the biphenylcarboxylic acid group) of amarogentin and amaroswerin (3)
Data needed	No data required
Specific observations	None
Classification and limits	Whole plant: category 4 (with limits on xanthones)
National/int. evaluation	Regulatory status in USA: chirata and chirata extract: CFR 172.510 (in alcoholic beverages only) (1)
Main references	(1) Fenaroli (1995) (2) Planta med. 38: 351-355 (1980) (3) Chem. Pharm. Bull. 34: 1663-1666 (1986) (4) Chem. Pharm. Bull. 32: 2290-2295 (1984a) (5) J. Pharm. Sci. 62 (6): 926-930 (6) Hager's Handbuch der Pharmazeutischen Praxis, 5th Ed., Haenseler R. et al. (Ed.), Springer Verlag, Berlin (1990) (7) J. Pharm. Sci. 65: 1547-1549 (1976) (8) Karrer W., Konstitution und Vorkommen der organischen Pflanzenstoffe, 3rd Ind. J. Chem. 31B: 70-71 (1992)

(9) Ind. J. Pharm. Sci. 44: 36 (1982)
(10) Leung (1996)
(11) J. Nat. Prod. 59: 27-29 (1996)
(12) Ind. J. Pharmacol. 27: 37-39 (1995)
(13) Drugs Exptl. Clin. Res. XIX (2): 69-73 (1993)
(14) Chem. Pharm. Bull. 32: 4942-4945 (1984b)

Data bases used Medline (1966-98)
Embase (1980-98)
Biological Abstracts (1989-98)
CC Life (2/97-2/98)
Keywords: Swertia chirata, Chirata, bitter stick

Triticum aestivum L. emend. Fiori et Paol.

CE No.	518 (2112)
Steinmetz No.	
FEMA No.	-
Order	Graminales
Family	Gramineae
Name	E Wheat
	F Amidon de blé, froment
	D Saatweizen, gemeiner Weizen
	I Grano tenero, frumento
	SP Trigo, extractos
Synonyms	Triticum vulgare Vill., T. sativum Lamk., T. durum Desf., T. hybernum L., T. turgidum L.
Parts used	Seed, oil
Important constituents	Seed: 82-85% starch, ca. 0.1% gluten, 15% water; oil: 45% oilic acid, 40% linolic acid, 8% palmitinic acid, 3-4% stearinic acid, 2-3% phytosterines, 2% lecithin, vitamins (1), various tocopherols and ubiquinone-9 (2)
Active principles	Not known
Other chemical components	Not known
Products in which used	Wheat or common corn is traditionally used as a food (starch, oil) (1). The distillate of the seeds is used in alc. beverages
Level of use	Seeds distillate: alcoholic beverages 12-16700 ppm (IOFI 1998)
Preparation	Distillate
Main toxicological data	No relevant data found
Data needed	No data required
Specific observations	Wheat is used as a foodstuff
Classification and limits	**Seeds, seed oil: category 1**
National/int. evaluation	None
Main references	(1) Hoppe H.A., Drogenkunde, 8th Ed., de Gruyter, Berlin (1975)
	(2) Biochem. J., 108, 465-473 (1968)
Data bases used	Medline (1966-98)
	Embase (1980-98)
	Biological Abstracts (1989-98)
	CC Life (2/97-2/98)
	Keywords: Triticum aestivum, Triticum vulgare, Triticum sativum

Urtica dioica L.

CE No.	468
Steinmetz No.	1176
FEMA No.	
GROUP	A11 07005
Order	Urticales
Family	Urticaceae
Name	E Common nettle, stinging nettle
	F Grande ortie
	D Brennessel, grosse Brennessel
	I Orticore
	SP Ortiga mayor
Synonyms	-
Parts used	Herb, leaves. Note: root no known flavour use according to IOFI (1)
Important constituents	Essential oil: methyl heptenone, citral, linalool
	Herb/Leaves: chlorophyll (rich source); carotenoids; phenolic acids-ferulic, caffeic, sinapic acids; Esculetin (2,3)
	Stinging hairs: acetyl choline (50mM) (4), choline, histamine, serotonin
	Plant: organic acids including quinic acid, glycollic acid, allantoic acid
Active principles	Not known
Other chemical components	Not known
Products in which used	Beverages (5,6,7)
	Sausages (8,9)
Level of use	Infusion 4% w/w in beverage (Russia) (7); 0.073% in sausage (Germany) (9). Young nettle leaves have been used in salads (10). Leaves are often used in soup. Herbs and leaves used in beverages at 0.2-5 g/L and in food at 0.1-100 g/kg (1)
Preparation	Extract, infusion (of herb and leaves)
Main toxicological data	Aqueous extract: LD50 (Mice i.p) 3.6 g/kg bw (11); Mice i.v.) 1.7 g/kg bw (12)
	Infusion: LD50 (Mice i.v) 1.99/kg bw; (rats, intragastric probe) 1.31 g/kgbw. Anti-coagulant due to coumarins (12)
	Stinging hairs: histamine, acetyl choline, serotonin-biogenic amines likely to be metabolised on oral ingestion
	Other components: chlorophyll, carotenes and xanthophylls permit-

	ted colourings. Methyl heptenone category B; citral, linalool, Blue Book: category A 4th Ed. Vol. I
Data needed	No data required
Specific observations	Herb and leaves used as a foodstuff
Classification and limits	**Herb: category 1** **Leaves: category 1**
National/int. evaluation	UK FACC (1976) Appendix 3; Herb, Leaves
Main references	(1) Iommunicanon from IOFI (1994). (2) Pharm. in uns. Zeit 12, (6) 181-186 (1983) (3) Buletinul Universit. Din Galati Fasc. 6,5 59-63 (1982) (4) Biochem. J. 132: 15-518 (1973) (5) Paten; Chem Abs. 103, 25, 213827 (1985) (6) Patent. Chemical Abstracts 103 (21) 177287. (1985) (7) Patent. FSTA Abstract 78-03-H0297. (1977) (8) Patent. FSTA Abstract 74-11-S1445. (1974) (9) FSTA Abs. 85-06-SO144. (1984) (10) J. Sci Food Agric. 31, 1279-1286. (1980) (11) Plantes medicinales et Phytotherapie XX(3): 219-226 (1986) (12) Anal. Bromatol. XXXV-1 99-103 (1983)
Data bases used	Biosis (1973-97) Chemical Abstracts (1967-97) FSTA (1969-97) Toxline (1981-97) Toxlit, Toxnet (1981-89) Medline (1966-97) Embase (1974-97) FROSTI (1975-88) *Keywords:* Urtica dioica nettle

Vaccinium macrocarpon Ait.

CE No.	470
Steinmetz No.	1179
FEMA No.	-
Order	Ericales
Family	Ericaceae
Name	E (Large) cranberry
	F Canneberge à gros fruit
	D Großfrüchtige Moosbeere
	I Mirtillo varietá
	SP Arandano agrio
Synonyms	Oxycoccus macrocarpos Ait

Parts used	Fruit
Important constituents	Fruit: anthocyanins (peonidin-3-galactoside, cyanidin-3-galactoside, cyanidin-3-arabinoside and others) (1), terpene hydrocarbons (limonene, 0.12 ppm), alfa-terpineol, acetic acid, benzoic acid, benzyl alcohol (2), pectin (3)
Active principles	Not known
Other chemical components	Not known
Products in which used	Cranberries are seldom used in flavours but are popular for the preparation of sauces, jelly and juice (3)
Level of use	Beverages 4g/l; food – (IOFI 1994)
Preparation	None
Main toxicological data	No relevant data found
Data needed	No data required
Specific observations	Cranberry is used as a foodstuff
Classification and limits	**Fruit: category 1**
National/int. evaluation	None
Main references	(1) J. Amer. Soc. Hort. Sci. 112(1), 100-104 (1987)
	(2) Z Lebensm. Unters. Forsch. 172, 365-367 (1981)
	(3) Heath, Source Book of Flavours (1981)
Data bases used	Chem Abstr. (1967-90)
	FSTA (1969-90)
	Biosis (1973-90)
	Toxline (1965-91)
	Medline (1966-90): Vaccinium, cranberr.
	Keywords: Vaccinium macrocarpom, Oxycoccus macrocarpom, cranberry

Vaccinium myrtillus L.

CE No.	469
Steinmetz No.	1178
FEMA No.	
Order	Ericales
Family	Ericaceae
Name	E Blueberry, bilberry
	F Myrtille
	D Heidelbeere, Blaubeere
	I Mirtillo nero
	SP Arandano
Synonyms	
Parts used	Fruit, leaves
Important constituents	Fruits: flavonoids (including quercitrin, rutin, isoquercitrin, myrtillin, anthocyan pigments); ursolic acid; coumaric acid; tannins (7%); volatile alcohols, terpene hydrocarbons and carbonyl compounds (1, 2, 3, 4) Leaves: flavonoids (including quercetin and those found in fruits); iridoid glucosides; phenol glucosides; vaccinin and neomyrtillin; tannins (0.8-6.7%); triterpenoids and alkaloids. Phenol glucosides (arbutin, 1.5 and hydrochinon, 1%) possibly present (2)
Active principles	Not known
Other chemical components	Not known
Products in which used	Fruits are used in the preparation of marmalades and jellies, used in dairy products or used in natural form. Derivatives are used for medical purposes (1)
Level of use	Annual use of fruit in Europe 13000 kg. Use level of leaves: beverages 3g/l (IOFI 1994)
Preparation	Derivatives used for medical purposes: fruits, fluid extract, tincture (20% in 20% alcohol); concentrated (6-8 fold) juice, leaves, infusion (3%), dried aqueous extract
Main toxicological data	Bilberries are traditionally used for treatment of gastro-intestinal infections, diarrhoea, circulatory disturbances and retinal degeneration (2). Leaves from V. myrtillus have suppressive effect on blood glucose levels, and are used in treatment of diabetic conditions (5). Chronic use of leaves for antidiabetic treatment was found to cause severe anemia in humans (2). In an experimental study, rats were given 1.5 g leaves/kg body weight and day. The animals showed

	serious anemia, ikterus and cachexia, and finally died. These symptoms have been claimed to be caused by the phenol glucosides in leaves from V. myrtillus
Data needed	Quantitative data on chemical components in leaves and, if necessary, toxicity data
Specific observations	Bilberries are used as a foodstuff
Classification and limits	**Fruit: category 1** **Leaves: category 5**
National/int. evaluation	None
Main references	(1) Fenaroli (1995) (2) List & Hörhammer (1967-80) 3 (3) J. Sci. Food Agric 34, 992-998 (1983) (4) Lebensm. Wiss. u. Technol. 2, 78-81 (1969) (5) Wichtl (1989)
Data bases used	FSTA (1969-90): Biosis (1973-90) Medline (1966-90) Toxline (1965-91) Chem. Abstr. (1969-90) *Keywords:* Vaccinium myrtill., bilberr.

Vaccinium uliginosum L.

CE No.	471
Steinmetz No.	1180
FEMA No.	
Order	Ericales
Family	Ericaceae
Name	E Bog bilberry, blueberry
	F Airelle des marais, myrtille de loup
	D Rauschbeere, Moosbeere
	I Mirtillo uliginoso
	SP Arandano negro
Synonyms	
Parts used	Fruits
Important constituents	Fruits: tannins (catechin), organic acids (benzoic acid)
Active principles	Not known
Other chemical components	Not known
Products in which used	Consumed as food item in some European countries (fruit)
Level of use	None
Preparation	None
Main toxicological data	After eating the berries symptoms of poisoning have occasionally been observed. No toxic substances have been found in V. uliginosum and therefore a fungus (Sclerotina megalospora) which sometimes parasitizes the berries has been suggested to be responsible for toxic symptoms (1)
Data needed	No data required
Specific observations	None
Classification and limits	Fruit: category 1
National/int. evaluation	None
Main references	(1) Frohne and Pfänder, A Colour Atlas of Poisonous Plants, London: Wolfe Publ. (1984)
	(2) List & Hörhammer (1967-80)
Data bases used	Med Line (1966-97)
	Chemical Abstracts (1967-97)
	Biosis (1976-97)
	Keywords: Latin and English names

Valeriana officinalis L.

CE No.	473
Steinmetz No.	1182
FEMA No.	Valerian root extract: 3099, Valerian root oil: 3100
Order	Dipsicales
Family	Valerianaceae
Name	E Valerian
	F Valériane
	D Gemeiner Baldrian
	I Valariana
	SP Valeriana
Synonyms	Valeriana alternifolia Ledeb., Valeriana sylvestris Grosch, Valeriana excelsa poir

Parts used	Roots
Important constituents	Iridoid compounds called valepotriates, which include valtranes, didrovaltranes and isovaltranes; valerosidatum, valeranone, alkaloids (actinidine, valerianine, valerine) (1,4)
Active principles	Not known
Other chemical components	Not known
Products in which used	Alcoholic and non-alcoholic beverages, condiment, relish, frozen dairy, breakfast cereals, dairy desserts, candy, baked goods, gelatin, puddings, meat, meat products (1,3)
Level of use	Valerian root extract: baked goods 94.26 ppm; breakfast cereals 3.0 ppm; frozen dairy 23.68 ppm; condiment, relish 41.0 ppm; soft candy 82.76 ppm; gelatin, puddings 52.65 ppm; non-alcoholic beverages 43.04 ppm; alcoholic beverages 96.06 ppm Valerian root oil: baked goods 16.68 ppm; frozen dairy 1.11 ppm; meat products 0.225 ppm; soft candy 5.26 ppm; gelatin, puddings 1.94 ppm; non-alcoholic beverages 0.95 ppm; alcoholic beverages 0.52 ppm; hard candy 0.01 ppm; chewing gum 390.09 ppm (3)
Preparation	Crude, extracts, and oil (valerian oil, kesso root oil, valerian root oil)(2,3)
Main toxicological data	Valerian has CNS-depressant activities and is reported to have antispasmodic and equalizing activities. The valepotriates have been reported to be mainly responsible for the CNS-depressant and antispasmidic effects in laboratory animals (1). The effects of a valepotriates mixture on mothers and progeny were evaluated in rats. A 30-day oral administration of valepotriates (a mixture containing

	80% dihydrovaltrate, 15% valtrate and 5% acevaltrate; daily doses were 6, 12, and 24 mg/kg; did not change the average length of estral cycle nor the number of estrous phases. No changes were found in fertility index. Internal examination of fetuses revealed an increased number of retarded ossifications in the two highest dose groups. No differences were seen in the development of the offspring. The valepotriates caused a hypothermisant effect after ip administration but not after oral administration. It was concluded by the authors that oral administration of valepotriates is inocuous to pregnant rats and their offspring (5)
Data needed	28-day study and mutagenicity studies on preparations. Study on CNS-effects
Specific observations	None
Classification and limits	**Roots: category 5**
National/int. evaluation	Valerian root extract CFR 172.510; valerian root oil CFR 172.510.
Main references	(1) Leung (1980) (2) Arctander (1960) (3) Fenaroli (1995) (4) Perfumer&Flavorist 10;98-102 (1985) (5) J. Ethnopharmacology 41:39-44 (1984)
Data bases used	Chemical Abstracts (1967-97) Biosis (1973-97) Toxline (1981-97) FSTA (1969-97) Medline (1966-97) *Keywords:* Latin names, methylpyrrylketone

Vanilla planifolia G.Jacks.

CE No.	474
Steinmetz No.	1185
FEMA No.	-
Order	Microspermae
Family	Orchidaceae
Name	E Vanilla
	F Vanille, Vanillier
	D Echte Vanille
	I Vaniglia
	SP
Synonyms	Vanilla aromatica Willd., Vanilla sativa Schiede, Vanilla fragans Salisb.

Parts used	Fruits
Important constituents	Vanillin, anisaldehyde, piperonal, eugenol (8)
Active principles	Not known
Other chemical components	Not known
Products in which used	Vanilla, vanilla extract, and vanilla oleoresins are widely used as flavour ingredients in most food products (1)
Level of use	Vanilla: baked goods 9642 ppm; breakfast cereal 460 ppm; frozen dairy 913.6 ppm; soft candy 1395 ppm; confection, frosting 1886 ppm; gelatin, puddings 434.2 ppm; non-alcoholic beverages 223.1 ppm; alcoholic beverages 2076 ppm; hard candy 293.0 ppm Vanilla extract: baked goods 3449 ppm; frozen dairy 3877 ppm; fruit juice 12.0 ppm; soft candy 2359 ppm; confection, frosting 4000 ppm; sweet sauce 437 ppm; gelatin, puddings 2732 ppm; non-alcoholic beverages 781.1 ppm; alcoholic beverages 311.4 ppm; initation dairy 400.0 ppm; hard candy 71.89 ppm Vanilla oleoresin: baked goods 454.1 ppm; frozen dairy 389.2 ppm; soft candy 352.2 ppm; gelatin, puddings 389.0 ppm; non-alcoholic beverages 239.6 ppm; alcoholic beverages 192.5 ppm; hard candy 25.93 ppm (2) Highest average maximum use level is about 9642 ppm (vanilla in baked goods) (1). Level of use in alcoholic beverages 10 g/l (IOFI 1994)
Preparation	Crude and extracts (1,2)
Main toxicological data	Vanillin: Long-term studies: carcino-negative (3). Mutagenicity: sister-chromatid exchanges (4 mM), metabolism (4). Increased

effect on sister-chromatid exchanges induced by chemical mutagens (5,6). JECFA ADI = 0-10 mg/kg bw/d (7)
Piperonal: Genotoxicity: positive in DNA repair test (Rec assay, Bacillus subtilles) (8)

Data needed	None
Specific observations	None
Classification and limits	**Fruit: category 1**
National/int. evaluation	GRAS §182.10, §182.20 and §169.3
Main references	(1) Leung (1980)
	(2) Fenaroli (1995)
	(3) Opdyke (1979)
	(4) Mut. Res. 190:221 (1987)
	(5) Mut. Res. 189:313 (1987)
	(6) Mut. Res. 191; 193 (1987)
	(7) WHO FAS 68.33/NMRS 44A-XI/78
	(8) Mut. Res. 101:127-140 (1982)
Data bases used	Chemical Abstracts (1967-97)
	Biosis (1973-97)
	FSTA (1969-97)
	Medline (1966-97)
	Toxline (1981-97)
	Keywords: Latin names, anisaldehyde, piperonal, vanillin

Vetiveria zizanoides (L.) Nash

CE No.	479
Steinmetz No.	1192
FEMA No.	-
Order	Graminales
Family	Gramineae
Name	E Vetiver
	F Vetiver Costus arabique
	D Vetiver
	I Vetiver
	SP Vetiver
Synonyms	-
Parts used	Rhizomes, roots (1)
Important constituents	The phenolic fraction of vetiver oil contains mainly zizanoic acid, smaller amounts of 4-vinylphenol, 4-vinyl-2-methoxyphenol and trans-isoeugenol among traces of other compounds (2). The alcoholic fraction contains a complex mixture of sesquiterpenes: α- and β- vetivon (10% of oil), khusimol (ca. 15% of oil), bicyclovetivenol, tricyclovetivenol, (-)-khusimone, (-)-epizizanoic acid, khusol, laevojunenol, norkhusinoloxide (3,4,5,6,7)
Active principles	Not known
Other chemical components	Not known
Products in which used	Vetiver alcohol aroma finds limited use in flavouring food substrates, mainly to reinforce the flavour of asparagus (1)
Level of use	Vetiver oil: average maximal use levels: alcoholic beverages 18 ppm, non-alcoholic beverages 8 ppm, ices 8 ppm, candy 12 ppm, baked goods 9 ppm, gelatin 9 ppm (IOFI 1996)
Preparation	Vetiver oil is obtained by steam distillation of sun-dried rootlets and rhizomes
Main toxicological data	No phototoxic effects are reported for vetiver oil (8)
Data needed	No data required
Specific observations	None
Classification and limits	Rhizomes, roots: category 2 Vetiver oil: category 2
National/int. evaluation	Vetiver: CFR 172.510; Vetiver oil: CFR 172.510

Main references

(1) Fenaroli (1995)
(2) Phytochemistry 21 (3): 793 (1982)
(3) Karrer W., Konstitution und Vorkommen der org. Pflanzenstoffe (1976)
(4) Karrer et al., Konstitution und Vorkommen der org. Pflanzenstoffe Ergänzungsband 1 (1977)
(5) Can. J. Chem. 60: 1081-1091 (1982)
(6) Tetrahedron 41 (16): 3387-3390 (1985)
(7) Bauer K. et al., Common Fragrance and Flavor Materials, p. 180 (1990)
(8) Food Cosmetic. Toxicol. 12: 1013 (1974)

Data bases used

Biological abstracts (1989-95)
Biosis (1979-95)
Chemical abstracts (1967-95)
Keywords: Vetiveria zizanoides

Viola odorata L.

CE No.	482
Steinmetz No.	1201
FEMA No.	Leaves absolute: 3110
Order	Violales
Family	Violaceae
Name	E Sweet Violet, Blue violet
	F Violette odorante
	D Duftveilchen, wohlriechendes Veilchen
	I Violetta, viola mammola
	SP Violeta
Synonyms	-
Parts used	Flowers, leaves (1)
Important constituents	Flower extract: (2E,6Z) – nonadienal, (2E-6Z) – nonadienol, hexanol, heptenol, an octadienol, benzyl alcohol, eugenol, 2-decanone, isoborneol, zingiberene, β-curcumene, dihydro-α-ionone, dihydro-β-ionone, α-ionone, β-ionone, vanillin and vitamin C (0.025% on a dry weight basis) (2,3). Leaf extract: (2E, 6Z) -nonadienal, (2E-6Z)-nonadienal, hexanol, 2-octenol, benzyl alchohol, an octenol, a hexenol, (Z)-4-methyl-2-hexenol, 4-isopropyl-2-pentenol and eugenol (2)
Active principles	Not known
Other chemical components	Not known
Products in which used	Leaf absolute used in food. Candied flowers used by confectioners (4). Flowers used in liqueurs, and other alcoholic beverages
Level of use	In the USA, leaf absolute: baked goods (16.64 ppm), frozen dairy (7.42 ppm), soft candy (6.68 ppm), gelatin, puddings (6.80 ppm), non-alcoholic beverages (2.58 ppm), alcoholic beverages (0.84ppm), hard candy (2.67 ppm), chewing gum (7.50 ppm) (1)
Preparation	Tincture (20% in 60% ethanol), concrete (flowers, leaves), infusion (5%), fluid extract and leaves absolute (FEMA No. 3110) (1)
Main toxicological data	Seeds are diuretic, vomitory and purgative. Alcoholic extract of leaves has a significant antipyretic effect in rabbits. Oral doses of up to 1.6g/kg tolerated well by rabbits (7 day study). No mortality occurred and no obvious side-effects recorded (5). A diuretic effect has also been reported (6). Aqueous extract of flowers possesses some antibiotic activity (3). Reported to have anti-inflammatory effects in acute and subacute (8-10-day) experiments on rats (7)

Data needed	Quantitative data on chemical components in leaves and, if necessary, 28-day oral study and mutagenicity studies
Specific observations	None
Classification and limits	**Flowers: category 1** **Leaves: category 5**
National/int. evaluation	UK FACC (1976) Appendix 2. Leaves absolute FEMA No.. 3110 Flowers and leaves GRAS (II)
Main references	(1) Fenaroli (1995) (2) Perfumer&Flavorist 19 (1), 33 (3) Zaida (1984), Pakistan J. Sci. Ind. Res. 27, 5 (4) Tanaka (1976) (5) Khattak et al. (1985), J. Ethnopharm, 14 (6) Rebuelta (1983), Plantes Med et Phytotherapie, XVII, 4, 215-21 (7) Fd. Chem. Tox. (1976) 14, 893
Data bases used	Biosis (1973-90) Chemical Abstracts (1967-90) FSTA (1969-90) Toxline, Toxlit, Toxnet (1981-90) Toxlit65 (1965-90) Embase (1974-90) *Keywords:* Viola odorata, Sweet violet

Viola tricolor L.

CE No.	483
Steinmetz No.	1202
FEMA No.	-
Order	Violales
Family	Violaceae
Name	E Wild pansy, Heartsease
	F Pensée sauvage
	D Stiefmütterchen
	I Viola tricolore
	SP Pensiamento
Synonyms	Viola arvensis Murr., Viola tricolor arvensis L.
Parts used	Flowers, herb (1)
Important constituents	Flowers: 10 mg/g dry flowers carotenoids (mainly violaxanthin 87%); anthocyanins (mainly violanin); rutin; methyl salicylate (trace)(1)
	Herb: Flavonoids (0.4% whole plant) possibly including rutin and quercetin; salicylic compounds; phenolic acids; sterols (1)
Active principles	Not known
Other chemical components	Not known
Products in which used	Alcoholic drinks, pastries, confectionery (1)
Level of use	Aromatised wines max 0.2 g/l; liqueurs max 0.4 g/l; alcoholic beverages 0.4 g/l (1)
Preparation	Infusion, extract, syrup (from herb)
Main toxicological data	Whole plant – no toxicological data. Individual components – very little toxicity data available but many components of available flowers are permitted as food colourants (2,3). Quercetin is mutagenic, carcinogenicity data inconclusive (4)
Data needed	Level of use in patisserie and confectionery. More quantitative chemical information on components present in herb or flower preparations and, if necessary, 28-day oral study and study on mutagenicity studies
Specific observations	None
Classification and limits	**Flowers: category 5**
	Herb: category 5

National/int. evaluation	UK FACC (1976)Appendix 3 USA, Pansy: FDA 121.1163 (in alcoholic beverages only)
Main references	(1) Comitato per lo studio delle Bevande Alcohooliche Aromatizzate, Unpublished report (1975) (2) Handbook Int. Food Reg. Toxicol. Vol 2 MTP Press (1981) (3) EC Directive 94/36/EC (4) Prog. Clin Biol. Res. 206:33-43 (1986)
Data bases used	Biosis (1973-97) Chemical Abstracts (1967-97) Toxlit, Toxnet (1981-97) Embase (1974-97) BFMIRA/FROSTI (1975-97) *Keywords:* Viola tricolor, Viola, Heartsease, Wild Pansy

Viverra zibetha Schreber

CE No.	3006
Steinmetz No.	Animal
FEMA No.	Absolute: 2319; Civetone: 3425; Skatole: 3019
Order	Carnivora
Family	Viverridae
Name	E Large Indian civet, Zibeth, Zibet
	F Civette
	D Zibetkatze
	I
	SP
Synonyms	Viverra civetta also called Civet but a different species, the African civet
Parts used	Glandular secretion of male and female civet cats, a yellow/brown paste (1)
Important constituents	Derivatives of the glandular secretion contain: skatole (3-methylindole) (1% of secretion), civetone, civetol, cis-6-methyltetrahydropyran-2-yl acetic acid. Several macrocyclic ketones (2,3,4,5)
Active principles	Not known
Other chemical components	Not known
Products in which used	Civet absolute is used in food as a flavouring
Level of use	In USA for civet abslolute: baked goods (13.10 ppm), frozen dairy (6.01 ppm), soft candy (14.16 ppm), gelatin, puddings (19.25 ppm), non-alcoholic beverages (2.11 ppm), alcoholic beverages (5.71 ppm), hard candy (0.25 ppm), chewing gum 1.10 ppm)(1). In Europe: foods (1.14 ppm), beverages (1-6 ppm) (6)
Preparation	Civet absolute, essence, tincture solvent extract
Main toxicological data	LD_{50} (oral, rat) civet abslolute > 5g/kg (2). Skatole lung toxicant in ruminants (3,7) due to methylene-imine metabolite (8). In vitro binding of metabolite to lung proteins in goat > rodents and man; in goats the binding to lung > to liver; in rodents and man binding to liver > to lung (7). In mice, 10 mg/kg bw i.p. for 15 days caused liver, spleen, kidney and lung lesions (9). Only acute oral studies available – LD_{50} rat 3.45 g/kg; mice 200 mg/kg wich gave lung congestion but no emphysema, whereas 100 mg/kg did not result in lung effects. In goat 300 mg/kg gave lung oedema and death (3). Skatole category 6 Blue Book 4th Ed. Volume I.

Specific observations	None
Data needed	Quantitative information on chemical components of civet preparations; 28-day toxicity study by oral route and in vivo metabolism study in rodents
Classification and limits	**Civet: category 5**
National/ int. evaluation	UK FACC (1976) Appendix 2 FEMA 2319 Civet absolute FEMA 3425 Civetone FEMA 3019 Skatole
Main references	(1) Fenaroli (1975) (2) Food Cosmet. Tox 12: 863 (1974) (3) Food Cosmet. Tox 14: 727 (1976) (4) Helv Chem. Acta 62, 4, p 1096-7 (1979) (5) Fenxi Hauxue Vol 11, p 781-3 (6) IOFI (1994) (7) Drug Metab. Disp. 19: 977 (1991) (8) Chem. Res. Toxicol. 5: 713 (1992) (9) Jap. J. Tuberc. 9: 65 (1961)
Data bases used	Biosis (1979-97) Frosti (1981-89) Chemical abstracts (1967-97) Toxlit (1981-89) FSTA (1969-97) Toxnet(1981-89) Medline (1966-97) Embase(1974-97) Toxline (1990-97) *Keywords:* Viverra zibetha, V. civetta, Civet, Civet absolute, Civet essence, 3-methyl indole, civetone

Zea mays L.

CE No.	488
Steinmetz No.	-
FEMA No.	Corn silk extract and oil: 2235
Order	Graminales
Family	Gramineae
Name	**E** Corn, maize
	F Maïs
	D Mais
	I Granoturco, Frumentone
	SP Maiz
Synonyms	Zea alba Mill., Zea americana Mill., Zea vulgaris Mill.

Parts used	Corn silk (stigmata maydis; maize female inflorescence or silk of maize)
Important constituents	Corn silk: rich in potassium salts; other constituents are ca. 12% polyphenoles (tannins), flavonoids, lipids (about 2% fatty oil), 3.8% gums, 2.7% resin, 3.2% saponines, 1.2% bitter glycosides, possibly up to 0.85% alkaloids, vitamins C and K, sitosterol, stigmasterol, plant acids, anthocyanins, reducing sugars, glucides, 0.1-0.2% essential oil with ca. 18% carvacrol (1) and other terpenes, mucilage (1,2,3,4)
Active principles	Not known
Other chemical components	Carvacrol
Products in which used	Corn silk is used as a flavouring in baked goods, frozen dairy, soft candy, gelatin and puddings, non-alcoholic and alcoholic beverages (5)
Level of use	Corn silk: baked goods 26.4 ppm, frozen dairy 10.9 ppm, soft candy 16.7 ppm, gelatin, puddings 2.7 ppm, non-alcoholic beverages 21.6 ppm, alcoholic beverages 0.2 ppm (5)
Preparation	Corn silk, corn silk extract and corn silk oil (5)
Main toxicological data	Corn silk is a well-known traditional remedy used as infusion (urinary tract infection, kidney stones, edema and diabetes). It has probably a mild diuretic effect by potassium salts. An immunostimulating effect has been shown (6). The effective dose in rabbits for diuretic, hypoglycemic, and hypotensive activities was 1.5 mg/kg b.w. (2)
Data needed	No data required

Specific observations	None
Classification and limits	**Corn silk: category 4 (with limits on carvacrol)** **Corn silk oil: category 4 (with limits on carvacrol)**
National/int. evaluation	Corn silk: GRAS (182.20) (Leung). FEMA No.. 2235; corn silk extract and oil: CFR 184.1262 (specification under development) (5)
Main references	(1) Hoppe H.A., Drogenkunde, 8th edition, de Gruyter, Berlin, New York (1975) (2) Leung (1996) (3) Wichtl M., Teedrogen. 2nd Ed., WVG, Stuttgart (1989) (4) Ageel A. et al., Plants used in Saudi Folk Medicine, King Saud Univ. Press, Riyadh (1987) (5) Fenaroli (1995) (6) ICMR Annals, 7, 37-49 (1987)
Data bases used	Medline (1966-98) Embase (1980-98) Biological Abstracts (1989-98) C Life (2/97-2/98) *Keywords:* Zea mays, corn, maize, stigma(ta) maidis/maydis, corn silk

Alphabetical index of the natural sources of flavourings

CoE No.	Latin name	Page
1	Abelmoschus moschatus Moench.	31
11	Acer saccharum Marsh.	35
12	Achillea millefolium L.	37
15	Aframomun melegueta K. Schum.	39
19	Agropyron repens (L.) Beauv.	41
264	Aloysia triphylla (L'Herit.) Britt.	43
33	Amyris balsamifera L.	45
56	Angelica archangelica L.	47
2008	Annona cherimola Mill.	51
46	Annona squamosa L.	53
49	Anthoxanthum odoratum L.	55
50	Anthriscus cerefolium (L.) Hoffm.	57
60	Artemisia abrotanum L.	59
61	Artemisia absinthium L.	61
64	Artemisia dracunculus L.	65
66	Artemisia spicata Wulf.	69
68	Artemisia umbelliformis Lam.	71
69	Artemisia pallens	73
70	Artemisia pontica L.	75
71	Artemisia vallesiaca Lam.	77
72	Artemisia vulgaris L.	81
2011	Artemisia herba-alba Asso.	85
78	Aspidosperma quebracho-blanco Schlechtend.	91
86	Berberis vulgaris L.	93
91	Boronia megastigma Nees ex Bartl.	95
93	Boswellia sacra Fleckiger	97
236	Bursera ssp.	99
103 b	Cananga odorata Hook. fil. et Thomson f. macrophylla	101
109	Carica papaya L.	103
112	Carum carvi L.	105
3002	Castor fiber L.	107
183	Centaurium erythraea Rafn.	109
2027	Cinchona officinalis L.	113
128	Cinchona pubescens Vahl	115

CoE No.	Latin name	Page
134	Cistus ladanifer L.	119
141	Citrus aurantiifolia (Christm.) Swingle	123
136 a	Citrus aurantium L. ssp. Aurantium L., rind	125
136 b	Citrus aurantium L. ssp. Aurantium L., flower	127
136 c	Citrus aurantium L. ssp. Aurantium L., leaf and twig	129
137	Citrus auriantum L. ssp. Bergamia (Risso&Poit.)	131
138	Citrus auriantum L. var. myrtifolia Ker-Gawl.	133
2032	Citrus japonica Thum.	135
139a	Citrus limon (L.) Burm. Rind	137
139b	Citrus limon (L.) Burm. leaf and twig	139
2035	Citrus medica L. var. medica	141
142	Citrus reticulata Blanco	143
2039	Citrus reticulata Blanco var. deliciosa H.H.Hu	145
2031	Citrus reticulata Blanco var. unshiu (Marco.) H.H.Hu	147
143	Citrus sinensis (L.) Pers.	149
140	Citrus x paradisi Macfad.	151
147	Cocos nucifera L.	153
149	Cola acuminata (P.Beauv.) Schott&Endl.	155
2041	Cola nitida (Vent.) Schott&Endl.	157
155	Corylus avellana L.	159
163	Curcuma longa L.	161
38	Cymbopogon citratus (DC.) Stapf	163
2045	Cymbopogon flexuosus (Nees ex Steud.) W. Wats.	165
40	Cymbopogon martinii (Roxb.) W.Wats. var. martinii	167
39	Cymbopogon nardus (L.) W.Wats.	169
2046A	Cymbopogon winterianus Jowitt	171
185	Eucalyptus globulus Labill.	173
194	Evernia prunastri (L.) Ach.	177
196	Ferula assa-foetida L.	179
197	Ferula gummosa Boiss.	181
2032	Fortunella japonica (Thunb.) Swingle	185
77	Galium odoratum (L.) Scop.	187
210	Gardenia jasminoides Ellis	191
214	Gentiana lutea L.	195
230	Hierochloe odorata L.	199
232	Hordeum vulgare L.	201
234	Hypericum perforatum L.	203
238	Illicium verum Hook.	205

Alphabetical index of the natural sources of flavourings

CoE No.	Latin name	Page
245	Jasminum grandiflorum L.	207
246	Jasminum officinale L.	209
255	Laurus nobilis L.	211
270	Mangifera indica L.	213
272	Marsdenia cundurango Rchb. f.	215
2076	Murraya koenigii (L.) Spreng.	217
294	Musa L. species, Musa sapientum L.	219
309	Olea europaea L.	221
345	Piper cubeba L.	223
352	Plantago lanceolata L.	225
361	Populus nigra L.	227
388	Quercus alba L.	229
2100	Saccharum officinarum L.	231
417	Sambucus nigra L.	233
420	Santalum album L.	235
427	Schinus molle L.	237
440	Swertia chirata Buch.-Ham. ex Wall.	239
2112	Triticum aestivum L. emend. Fiori et Paol.	243
468	Urtica dioica L.	245
470	Vaccinium macrocarpon Ait.	247
469	Vaccinium myrtillus L.	249
471	Vaccinium uliginosum L.	251
473	Valeriana officinalis L.	253
474	Vanilla planifolia G.Jacks.	255
479	Vetiveria zizanoides (L.) Nash	257
482	Viola odorata L.	259
483	Viola tricolor L.	261
3006	Viverra zibetha Schreber	263
488	Zea mays L.	265

List of national delegates of the Committee of Experts on Flavouring Substances

Austria

Dipl. Ing. R Bernhart
Bundesanstalt für Lebensmitteluntersuchung und -forschung
Kinderspitalgasse 15
A-1090 Wien
Tel: 43 1 40490-27864
E-mail: baluf@baluf.gv.at
Fax: 43 1 40490 27865

Belgium

Dr M-P Delcour-Firquet
Chef de travaux
Institut Scientifique de Santé Publique – Louis Pasteur
14 rue J. Wytsman
B-1050 Bruxelles
Tel: 32 2/642 51 28
E mail: m.p.delcour@iph.fgov.be
Fax: 32 2/642 52 24

Denmark

Dr J Gry
Scientific Adviser, Ph. D.
Institute of Food Safety and Toxicology
Mørkhøj Bygade 19
DK-2860 Søborg
Tel: 45 33/95 60 00
E-mail: jg@VFD.DK
Fax: 45 33/95 66 99

Dr H Frandsen
Institute of Food Safety and Toxicology
Danish Veterinary and Food Administration
Mørkhøj Bygade 19
DK-2860 Søborg
Tel: 45 33 95 60 00
E-mail: HF@VFD.DK
Fax: 45 33 95 66 99

Mrs H Juel Christoffersen
Institute of Food Safety and Toxicology
Danish Veterinary and Food Administration
Mørkhøj Bygade 19
DK-2860 Søborg
Tel: 45 33 95 60 00
E-mail: HJC@VFD.DK
Fax: 45 33 95 66 99

France

Prof. P L Nguyen
37 avenue Voltaire
F-91140 Bures-sur-Yvette
Tel: 33 (1) 69 07 43 43

M B Lacourt
Direction générale de la Concurrence de la
Consommation et de la répression des fraudes
Bureau C2
59, boulevard Vincent Auriol
F-75703 Paris Cedex 13
Tel: 33 1 44 97 32 03
E-mail: bruno.lacourt@dgccrf.finances.gouv.fr
Fax: 33 1 44 97 30 43

Germany

Dr G Matthiaschk
Federal Institute for Health Protection of Consumers and Veterinary Medicine
Thielallee 88 – 92
D-14195 Berlin
Tel: 49 188 84 12 34 16
Fax: 49 188 84 12 37 63

Italy

Dr M De Vincenzi
Dirigente di Ricerca
Istituto Superiore di Sanità
Viale Regina Elena 299
I-00161 Rome
Tel: 39 6/4990 25 26/2598
Fax: 39 6/49387149

Spain

Dr P Burdaspal
Head – Chemical Area
Centro Nacional de Alimentacion (Instituto de Salud Carlos III)
SP-28220 Majadahonda (Madrid)
Tel: 34 91 50 97 931
E-mail: pburdas@isciii.es
Fax: 34 91 509 79 26

Dr J Bustos
Centro Nacional de Alimentacion – National Institute "Carlos III"
Ministerio de Sanidad y Consumo
Ctra. Hajadahonda -Pozuelo Km.2
SP-28220 Majadahonda – Madrid
Tel: 34 9 1/5097900
Fax: 34 9 1/5097926

Ms L Suarez González
Jefe de Sección
Ministerio de Sanidad y Consumo
Subdirección General de Higiene de los Alimentos
Paseo del Prado 18-20
SP-28014 Madrid
Tel: 34 91 5961972
E-mail: lsuarez@msc. se
Fax: 34 91 5964409

Suède

Ms U Beckman Sundh
Toxicologist -National Food Administration
Box 622
S-75126 Uppsala
Tel: 46 18 175 755
E-mail: USBE@SLV.SE
Fax: 46 18 10 5848

Switzerland

Dr J Amberg-Müller
Bundesamt für Gesundheit – Fachstelle Toxikologie
c/o Institut für Veterinärpharmakologie, und-Toxikologie
Winterthurerstrasse 260
CH-8057 Zürich
Tel: 41 1 635 87 78
E mail: judith.amberg@bag.admin.ch
Fax: 41 1 635 89 40

United Kingdom

Dr A M Davies
Senior Scientific Officer
Ministry of Agriculture Fisheries and Food
Room 232 Ergon House c/o Nobel House
17 Smith Square
GB-London SW1P 3JR
Tel: 44 171 238 6217
E-mail: a.davies@fsci.maff.gov.uk
Fax: 44 171 238 6263

Commission of the EU

Dr G A Schreiber
Commission Européenne
DG Enterprises/E1 AN88/3154
200 rue de la Loi
B-1049 Bruxelles
Tel: 32 2 295 65 40
E-mail: georg.schreiber@cec.eu.int
Fax: 32 2 295 17 35

Secretary of the Committee of Experts on Flavouring Substances

Dr Peter Baum
Partial Agreement Department in the Social and Public Health field
Council of Europe
Avenue de l'Europe
F-67000 Strasbourg
Tel: +33 (0)3 88 41 21 76
E-mail: peter.baum@coe.int
Fax: +33 (0)3 88 41 27 32

The members of the Committee of experts were greatly saddened to learn of the death of Mme G. Clair, Maître de Conférences, Département de Pharmacognosie, Faculté de Pharmacie, Université René Descartes, Paris.

Her active participation and involvement in the evaluation of the natural sources of flavourings contributed to the success of this report.

Council of Europe publications on flavouring substances

Flavouring substances and natural sources of flavourings, Bilingual 4th edition.
Part I: Chemically-defined flavouring substances, 1992,
ISBN 2-7160-0147-2

Resolution AP (85) 2 concerning the transmission of the flavour of smoke to food, 1985

Guidelines concerning the transmission of flavour of smoke to food, 1992,
ISBN 92-871-2191-5

Health aspects of using smoke flavours as food ingredients, 1992,
ISBN 92-871-2189-3

Guidelines for flavouring preparations produced by enzymatic or microbiological processes, 1994,
ISBN 92-871-2586-4

Guidelines for safety evaluation of thermal process flavourings, 1995,
ISBN 92-871-2811-1

Guidelines for flavouring preparations produced by plant tissue culture, 1998,
ISBN 92-871-3738-2

Sales agents for publications of the Council of Europe
Agents de vente des publications du Conseil de l'Europe

AUSTRALIA/AUSTRALIE
Hunter Publications, 58A, Gipps Street
AUS-3066 COLLINGWOOD, Victoria
Tel.: (61) 3 9417 5361
Fax: (61) 3 9419 7154
E-mail: Sales@hunter-pubs.com.au
http://www.hunter-pubs.com.au

AUSTRIA/AUTRICHE
Gerold und Co., Graben 31
A-1011 WIEN 1
Tel.: (43) 1 533 5014
Fax: (43) 1 533 5014 18
E-mail: buch@gerold.telecom.at
http://www.gerold.at

BELGIUM/BELGIQUE
La Librairie européenne SA
50, avenue A. Jonnart
B-1200 BRUXELLES 20
Tel.: (32) 2 734 0281
Fax: (32) 2 735 0860
E-mail: info@libeurop.be
http://www.libeurop.be

Jean de Lannoy
202, avenue du Roi
B-1190 BRUXELLES
Tel.: (32) 2 538 4308
Fax: (32) 2 538 0841
E-mail: jean.de.lannoy@euronet.be
http://www.jean-de-lannoy.be

CANADA
Renouf Publishing Company Limited
5369 Chemin Canotek Road
CDN-OTTAWA, Ontario, K1J 9J3
Tel.: (1) 613 745 2665
Fax: (1) 613 745 7660
E-mail: order.dept@renoufbooks.com
http://www.renoufbooks.com

**CZECH REPUBLIC/
RÉPUBLIQUE TCHÈQUE**
USIS, Publication Service
Havelkova 22
CZ-130 00 PRAHA 3
Tel./Fax: (420) 2 2423 1114

DENMARK/DANEMARK
Munksgaard
35 Norre Sogade, PO Box 173
DK-1005 KØBENHAVN K
Tel.: (45) 7 733 3333
Fax: (45) 7 733 3377
E-mail: direct@munksgaarddirect.dk
http://www.munksgaarddirect.dk

FINLAND/FINLANDE
Akateeminen Kirjakauppa
Keskuskatu 1, PO Box 218
FIN-00381 HELSINKI
Tel.: (358) 9 121 41
Fax: (358) 9 121 4450
E-mail: akatilaus@stockmann.fi
http://www.akatilaus.akateeminen.com

FRANCE
C.I.D.
131 boulevard Saint-Michel
F-75005 PARIS
Tel.: (33) 01 43 54 47 15
Fax: (33) 01 43 54 80 73
E-mail: cid@msh-paris.fr

GERMANY/ALLEMAGNE
UNO Verlag
Proppelsdorfer Allee 55
D-53115 BONN
Tel.: (49) 2 28 94 90 231
Fax: (49) 2 28 21 74 92
E-mail: unoverlag@aol.com
http://www.uno-verlag.de

GREECE/GRÈCE
Librairie Kauffmann
Mavrokordatou 9
GR-ATHINAI 106 78
Tel.: (30) 1 38 29 283
Fax: (30) 1 38 33 967

HUNGARY/HONGRIE
Euro Info Service
Hungexpo Europa Kozpont ter 1
H-1101 BUDAPEST
Tel.: (361) 264 8270
Fax: (361) 264 8271
E-mail: euroinfo@euroinfo.hu
http://www.euroinfo.hu

ITALY/ITALIE
Libreria Commissionaria Sansoni
Via Duca di Calabria 1/1, CP 552
I-50125 FIRENZE
Tel.: (39) 556 4831
Fax: (39) 556 41257
E-mail: licosa@licosa.com
http://www.licosa.com

NETHERLANDS/PAYS-BAS
De Lindeboom Internationale Publikaties
PO Box 202, MA de Ruyterstraat 20 A
NL-7480 AE HAAKSBERGEN
Tel.: (31) 53 574 0004
Fax: (31) 53 572 9296
E-mail: lindeboo@worldonline.nl
http://home-1-worldonline.nl/~lindeboo/

NORWAY/NORVÈGE
Akademika, A/S Universitetsbokhandel
PO Box 84, Blindern
N-0314 OSLO
Tel.: (47) 22 85 30 30
Fax: (47) 23 12 24 20

POLAND/POLOGNE
Główna Księgarnia Naukowa
im. B. Prusa
Krakowskie Przedmiescie 7
PL-00-068 WARSZAWA
Tel.: (48) 29 22 66
Fax: (48) 22 26 64 49
E-mail: inter@internews.com.pl
http://www.internews.com.pl

PORTUGAL
Livraria Portugal
Rua do Carmo, 70
P-1200 LISBOA
Tel.: (351) 13 47 49 82
Fax: (351) 13 47 02 64
E-mail: liv.portugal@mail.telepac.pt

SPAIN/ESPAGNE
Mundi-Prensa Libros SA
Castelló 37
E-28001 MADRID
Tel.: (34) 914 36 37 00
Fax: (34) 915 75 39 98
E-mail: libreria@mundiprensa.es
http://www.mundiprensa.com

SWITZERLAND/SUISSE
BERSY
Route d'Uvrier 15
CH-1958 LIVRIER/SION
Tel.: (41) 27 203 73 30
Fax: (41) 27 203 73 32
E-mail: bersy@freesurf.ch

UNITED KINGDOM/ROYAUME-UNI
TSO (formerly HMSO)
51 Nine Elms Lane
GB-LONDON SW8 5DR
Tel.: (44) 171 873 8372
Fax: (44) 171 873 8200
E-mail: customer.services@theso.co.uk
http://www.the-stationery-office.co.uk
http://www.itsofficial.net

**UNITED STATES and CANADA/
ÉTATS-UNIS et CANADA**
Manhattan Publishing Company
468 Albany Post Road, PO Box 850
CROTON-ON-HUDSON,
NY 10520, USA
Tel.: (1) 914 271 5194
Fax: (1) 914 271 5856
E-mail: Info@manhattanpublishing.com
http://www.manhattanpublishing.com

STRASBOURG
Librairie Kléber
Palais de l'Europe
F-67075 STRASBOURG Cedex
Fax: (33) 03 88 52 91 21

Council of Europe Publishing/Editions du Conseil de l'Europe
F-67075 Strasbourg Cedex
Tel.: (33) 03 88 41 25 81 – Fax: (33) 03 88 41 39 10
E-mail: publishing@coe.int – Web site: http://book.coe.fr